The Price Principles of the Grand Unified Theory

By

Douglas M. Price

ISBN: 1-4107-4907-X (e-book)
ISBN: 1-4107-4908-8 (Paperback)

Library of Congress Control Number: 2003094233

This book is printed on acid free paper.

Printed in the United States of America
Bloomington, IN

1st Books - rev. 12/01/03

MEL PRICE---MY FATHER---MY MENTOR

I shall never forget how he helped to shape my mind. This book wouldn't exist without his upbringing. He would say things to me, in my childhood, that stays in my mind. A few of the many quotes are:

WHEN YOU WAKE UP IN THE MORNING, TRY TO MAKE THE WORLD A BETTER PLACE THAN IT WAS WHEN YOU CAME INTO IT.

SEE IT LIKE IT IS AND SAY IT LIKE IT IS.

COUNT TEN BEFORE YOU SPEAK.

CALL A SPADE A SPADE.

BELIEVE LITTLE OF WHAT YOU HEAR AND ABOUT HALF WHAT YOU SEE.

YOU CAN JUDGE A PERSON BY THE COMPANY THAT HE KEEPS.

THE *BEST* WAY TO SAVE MONEY IS NOT TO SPEND IT.

WISH INTO ONE HAND & DO-DO IN THE OTHER HAND, AND SEE WHICH HAND GETS THE FULLEST.

IF YOU STEAL, STEAL BIG, BECAUSE THE PUNISHMENT FOR THE CRIME IS THE SAME FOR STEALING 35¢ AS IT IS FOR STEALING A MILLION DOLLARS.

THAT "CROCKHEAD" CAN NOT SEE BEYOND HIS OWN NOSE!

YOU TAKE A CLOSE LOOK INTO THE MIRROR, I MEAN A REAL CLOSE LOOK!

*---and many – many more. [que reposa]

DOUGLAS M. PRICE

DEDICATION

Special thanks to my good friend Nick Swenn for his assistance and support in the preparation, organization, and editing of my book. Nicks interest and 36 years experience in education provided me with motivation and encouragement to produce this book. He stated many times that he observed the strong interest and excitement shown by the educationally and intellectually gifted students in "science".

We believe this book will provide important additional knowledge and understanding for high school and college science students and all scientists and researchers the world over.

NICHOLAS SWENN
Graduated - Carnegie High School 1953
B.S.ED. - University of Pittsburgh 1957
M.ED - Duquesne University 1961
Post Graduate + Pre Doctoral Studies:
Duquesne University 1962-1968
West Virginia University 1963-1967
Pennsylvania State University 1968-1969

DAUGHTERS:
Stacy Williams- Medical Doctor
Jennifer Giotis- High school honors english teacher

FORWARD

Sub-atomic research that endeavors to reveal the nature and structure of elementary particles involves the use of large superconducting magnets that are placed about a huge ring in order to accelerate protons to reach great speeds. The huge ring device is known as a particle accelerator. It spends beams of protons in opposite directions causing collisions to occur. The collision causes the protons to break apart into smaller particles of mass, as is illustrated in the following diagram.

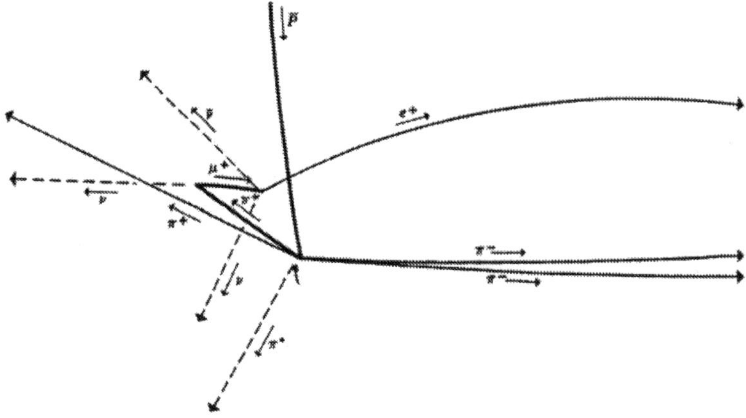

A research facility that is located in Europe, near Geneva, is planning a so called Super-Collider in order to collide Hadrons, which are larger than protons. I don't think that the sky-high costs of that endeavor will reveal the underlying structural details that we are looking for. We have already gathered enough data to, "figure out' the initial geometrical and mathematical manner that can reveal exactly how mass is formed at zero-space. That is what this book is about.

The drawing depicts the unknown activity that occurs during the dismantling of the products produced by the collision. Firstly, the effort of the many "Old Masters" of the past era, such as Aristotle-Newton-Franklin-Reimann-Einstein-Heisenberg etc., has provided us with enough of the aspects of the math and geometry concerning the nature of elementary particles so that we are able put together the puzzle and overcome the mystery that has eluded mankind from way back to the age of the cave dwellers.

Aristotle postulated that *everything* starts from one and progresses in growth in consecutive order, as 1-2-3-4-5-etc. I had to modify Aristotles postulate so that it begins at zero instead of one in order to account for space itself. Werner Heisenbergs Uncertainty Principle points out the fact that momentum and position cannot be detected (focused) in a system at the same time at the wee small realm of electrons. The formula is [$\Delta X \cong \Delta Y$]. X = momentum Y = position. Also, in order to overcome the Heisenberg dilemma, I reversed the timing sequence in a system to *come and go* in a way that it focuses momentum and position at the midpoint meeting place in the system. The simple math scheme can be illustrated as:

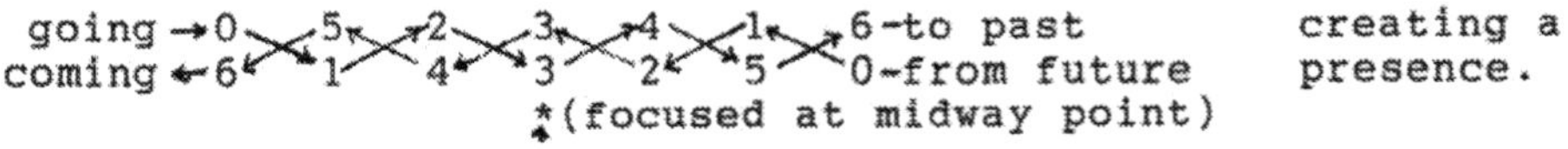

```
going →0  5  2  3  4  1  6-to past          creating a
coming ←6  1  4  3  2  5  0-from future      presence.
                     *(focused at midway point)
```

SECTION I

MOST OF THE CHARTINGS IN SECTION I ARE DIMENSIONAL DRAWINGS THAT ONLY SHOW LENGTH PLUS WIDTH WHICH IS A FLAT PORTRAYAL OF A PLANE GEOMETRY STYLE SCHEME. THE 3-D SCHEMES ILLUSTRATING LENGTH, WIDTH AND DEPTH AND 4-D (TIME) ARE DISPLAYED IN SECTION II.

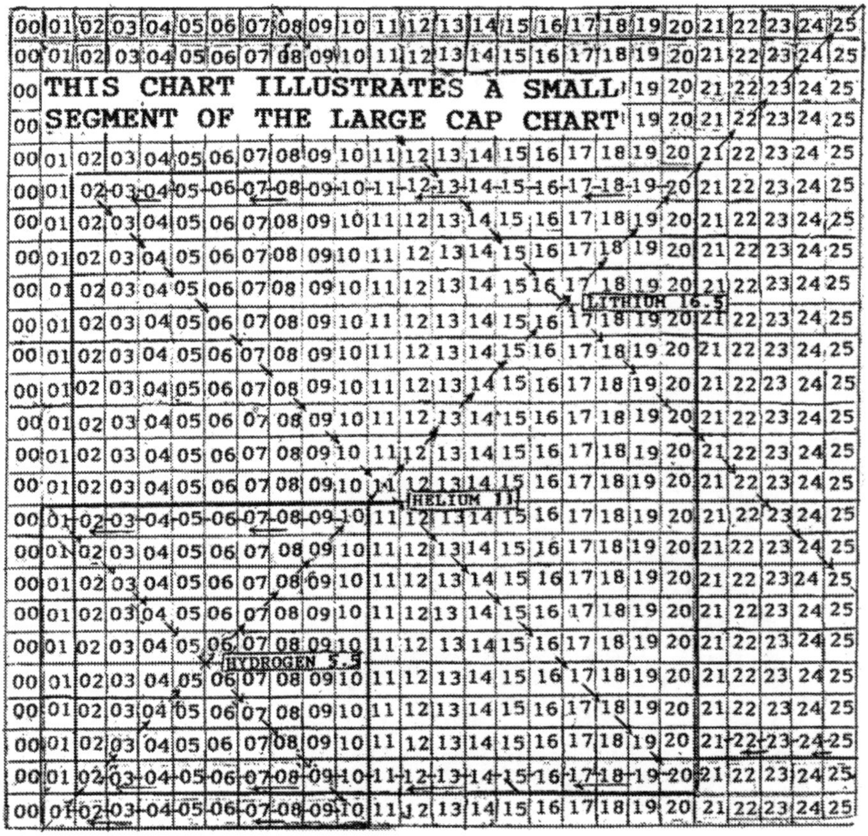

00	01	02	03	04	05	06	07	08	09	10	11	12	13	14	15	16	17	18	19	20	21	22	23	24	25
00	01	02	03	04	05	06	07	08	09	10	11	12	13	14	15	16	17	18	19	20	21	22	23	24	25
00	THIS CHART ILLUSTRATES A SMALL																		19	20	21	22	23	24	25
00	SEGMENT OF THE LARGE CAP CHART																		19	20	21	22	23	24	25
00	01	02	03	04	05	06	07	08	09	10	11	12	13	14	15	16	17	18	19	20	21	22	23	24	25
00	01	02	03	04	05	06	07	08	09	10	11	12	13	14	15	16	17	18	19	20	21	22	23	24	25
00	01	02	03	04	05	06	07	08	09	10	11	12	13	14	15	16	17	18	19	20	21	22	23	24	25
00	01	02	03	04	05	06	07	08	09	10	11	12	13	14	15	16	17	18	19	20	21	22	23	24	25
00	01	02	03	04	05	06	07	08	09	10	11	12	13	14	15	16	17	18	19	20	21	22	23	24	25
00	01	02	03	04	05	06	07	08	09	10	11	12	13	14	15	16	17	18	19	20	21	22	23	24	25
00	01	02	03	04	05	06	07	08	09	10	11	12	13	14	15	16	17	18	19	20	21	22	23	24	25
00	01	02	03	04	05	06	07	08	09	10	11	12	13	14	15	16	17	18	19	20	21	22	23	24	25
00	01	02	03	04	05	06	07	08	09	10	11	12	13	14	15	16	17	18	19	20	21	22	23	24	25
00	01	02	03	04	05	06	07	08	09	10	11	12	13	14	15	16	17	18	19	20	21	22	23	24	25
00	01	02	03	04	05	06	07	08	09	10	11	12	13	14	15	16	17	18	19	20	21	22	23	24	25
00	01	02	03	04	05	06	07	08	09	10	11	12	13	14	15	16	17	18	19	20	21	22	23	24	25
00	01	02	03	04	05	06	07	08	09	10	11	12	13	14	15	16	17	18	19	20	21	22	23	24	25
00	01	02	03	04	05	06	07	08	09	10	11	12	13	14	15	16	17	18	19	20	21	22	23	24	25
00	01	02	03	04	05	06	07	08	09	10	11	12	13	14	15	16	17	18	19	20	21	22	23	24	25
00	01	02	03	04	05	06	07	08	09	10	11	12	13	14	15	16	17	18	19	20	21	22	23	24	25
00	01	02	03	04	05	06	07	08	09	10	11	12	13	14	15	16	17	18	19	20	21	22	23	24	25
00	01	02	03	04	05	06	07	08	09	10	11	12	13	14	15	16	17	18	19	20	21	22	23	24	25
00	01	02	03	04	05	06	07	08	09	10	11	12	13	14	15	16	17	18	19	20	21	22	23	24	25
00	01	02	03	04	05	06	07	08	09	10	11	12	13	14	15	16	17	18	19	20	21	22	23	24	25
00	01	02	03	04	05	06	07	08	09	10	11	12	13	14	15	16	17	18	19	20	21	22	23	24	25

LITHIUM 16.5

HELIUM 11

HYDROGEN 5.5

ATOM CAPACITY CHART

Briefly, this chart illustrates the *potential* field that is capable of coming into existence when "odd" amounts of virtual momentum are consecutively combined within an all "evened" given area of Hilbert orthogonal (averaged) space. When this field acquires the capacity that houses 100 positions (a 10X10 area) it results in the potential positive+ outgoing field of a hydrogen atom (see lower left segment on the CAP chart). An electron's negative- inward coming field is housed in a 10X10 oppositely directed scheme. The 10X10 hydrogen atoms field is activated into being by the co-junction of an equal but oppositely directed coming (inward) 10X10 field.

0	01	02	03	04	05	06	07	08	09
0	01	02	03	04	05	06	07	08	09
0	01	02	03	04	05	06	07	08	09
0	01	02	03	04	05	06	07	08	09
0	01	02	03	04	05	06	07	08	09
0	01	02	03	04	05	06	07	08	09
0	01	02	03	04	05	06	07	08	09
0	01	02	03	04	05	06	07	08	09
0	01	02	03	04	05	06	07	08	09
0	01	02	03	04	05	06	07	08	09

This cap chart portrays the ***potential*** arrangement of atoms and their positron charge (+) configurations, which begin at the base of the field and end at the left top corner of the field. The positive charge is ***compressed*** from the right to left before it reaches the end of the field. THIS ALLOWS THE GOING/COMING (positive force) TO REVISE INTO A NEGATIVE (-) COMING/GOING force) that is inducted from the (far out) spaced electron and proceeds on down into the crossover mid point of the field, where the nucleus manipulates the proton-neutron functions.

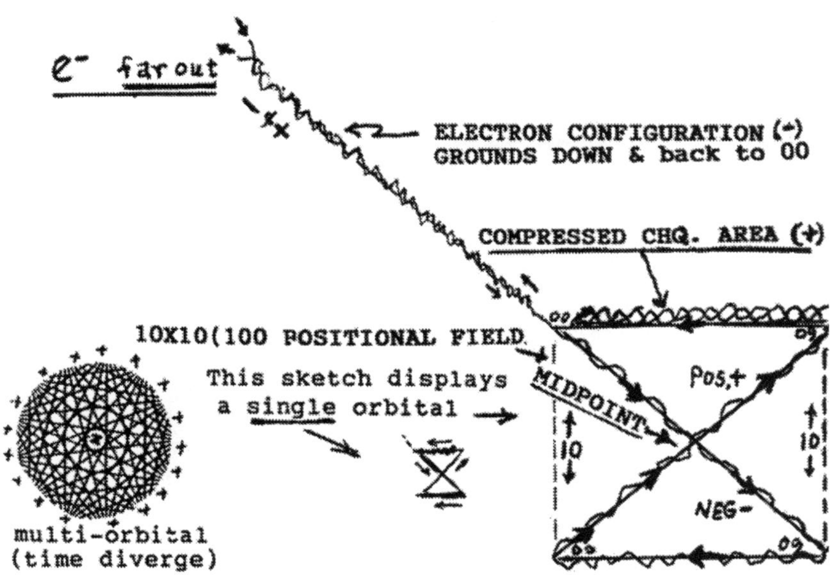

When the above illustrated "potential field is accelerated to its maximum capacity (99), the potential hydrogen space area becomes an atom, having a proton that manipulates at the place where the diagonal currents cross (+ and -) over at the midpoint. [Like forces repel each other. Unlike attract].

* THE MANIPULATIONS THAT OCCUR IN THE NUCLEUS (midpoint) WILL BE DESCRIBED AFTER THE FOLLOWING DIAGRAMS.

The following diagram may be used to better grasp the electrical field when it undergoes the coming/while/going current system. This field is also shown on the DNA starter chart displayed elsewhere in the *Heisenberg Zone Book.*

MOMENTUM *CHANGING* POSITIONS IN THE CHART, COMING FROM 0 WHILE GOING TO 99 BOTH WAYS

(This caption was placed below the charts in the original.)

This chart may be used to illustrate a *negative* electron field and by reversing the math scheme of the current, it becomes the field that houses a *positive* positron.

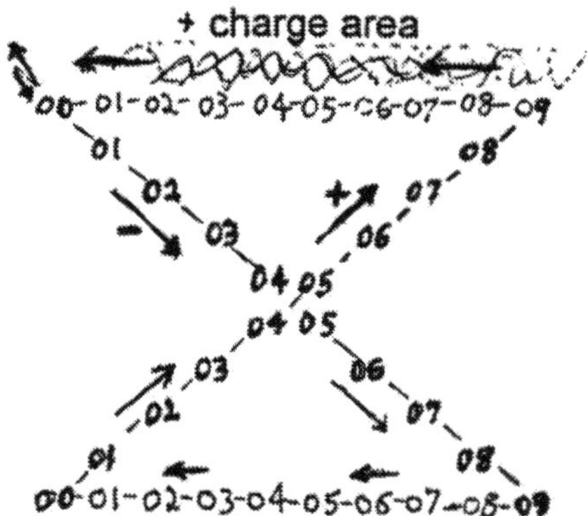

THE NEGATIVE CURRENT (-) OF *AN ELECTRON BOUND TO PROTON* (+) [WHEN ODD IS ADDED TO EVEN]
(This caption was placed below the diagram in the original.)

The potential electrical configuration that will occur in an even-odd consecutively arranged field sets up as is displayed in the above diagram of Hilbert orthogonal (averaged out) space. Odd adds motion into evenness. The diagram (above) depicts the + and – configuration that is in a hydrogen atom.

EINSTEIN – Time is constant

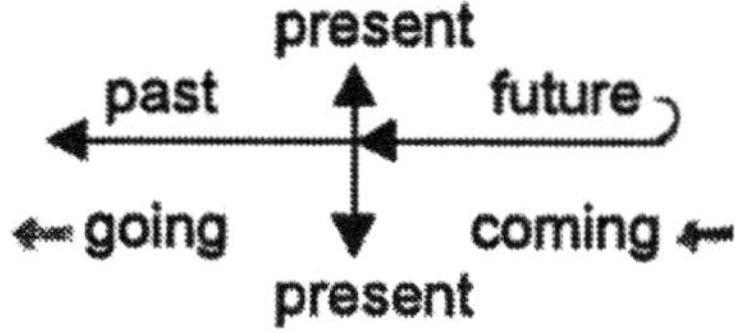

The time change from past to future – instantly (C^2). Thus, in this scheme, no presence of time is established in order to form an "existence" of a system in space.

The even amounts of momenta, on the above chart, are found anywhere and everywhere in space. The potential is there for the additional of "odd" amounts to activate a sub-atomic system that can be mathematically charted as such, because even and odd are required to define motion. (Continued on next page).

00	01	02	03	04	05	06	07	08	09
10	11							18	19
20		22					27		29
30			33			36			39
40				44	45				49
50				54	55				59
60			63			66			69
70		72					77		79
80	81							88	89
90	91	CONFIGURATION						89	

IN THE ABOVE CHART

A FREE ELECTRON CONFIGURATION

Compare this scheme to the top graph on page 4. The above-shown charts display the math scheme of a free electron (not bound to a proton). It maintains a stable "*presence*" in space. A *presence* is established by its timing sequence that uses some part time joined to some future time that forms a system.

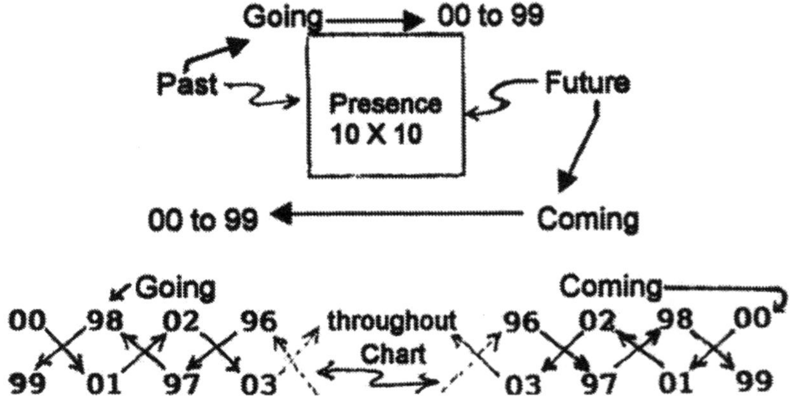

Thus it requires a *presence* of "time" in order for a subatomic particle to exist.

PAUL DIRAC suggested that an electron sort of disappears in and out of space (like coming-while-going). Paul employed a *simple* math scheme to describe the action as:

0	1
1	0

The actions performed by an electron are fully described in an earlier copyrighted article named "*The Price Principles of the Heisenberg Zone*".

The electron particle abounds in realistivic actions, relative to its angle of *function* of its position and momentum, see chart at top of page 8.

This space-transformation falls under Aristotle's postulates that "EVERYTHING BEGINS WITH ONE AND PROCEEDS IN GROWTH IN A CONSECUTIVE MANNER". He should have said "starts at zero and proceeds to grow in a consecutive manner".

The chart shown on page 4 of a 100 (10X10) position *potential* "Hilbert Space" field becomes activated when the field area is accelerated to be "filled" to its "total" capacity. In order to reach its full capacity, each position must contain an amount that totals 99.[0 through 99] is the total numbers necessary to fill 100 positions.

The "activated" field is the same field that surrounds a proton in a Hydrogen atom. The time diverge scheme, that is displayed on this capacity chart, depicts why it is very difficult to determine the exact position of an electron.

The electron orbitals, that surround the field of atoms are considered to be like a cloud. The electron is stable because it maintains the maximum amount that is able to be housed in each of the 100 positions in the 10X10 system. 99 is the maximum. It is arranged in the system starting across the top row as:

99 01 97 03 95

00 98 02 96 04

COMING *WHILE* GOING

The next –shown chart illustrates the *path* of an electron configuration and displays the 99 increase up and down diagonals. The diagonal path requires more distance than the horizontal. The difference is 5 compared to 7. The perimeter of a square, compared to its diagonal, is 20/7 (a endless repeating sequence as: 20/7 = 2.857142857142857 --- etc.)

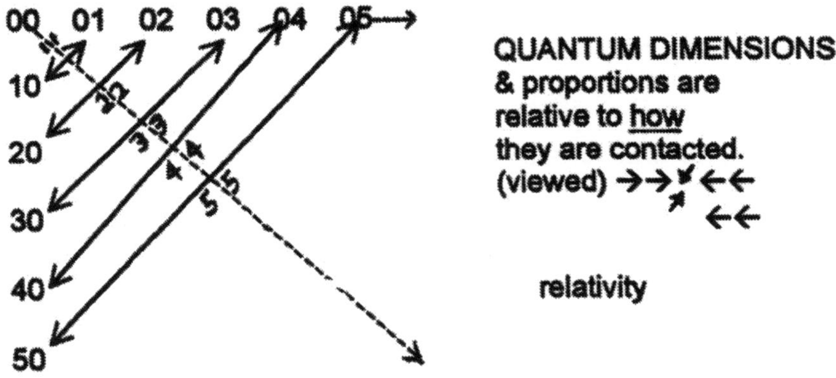

(taken from the free electron chart displayed on page 6.)

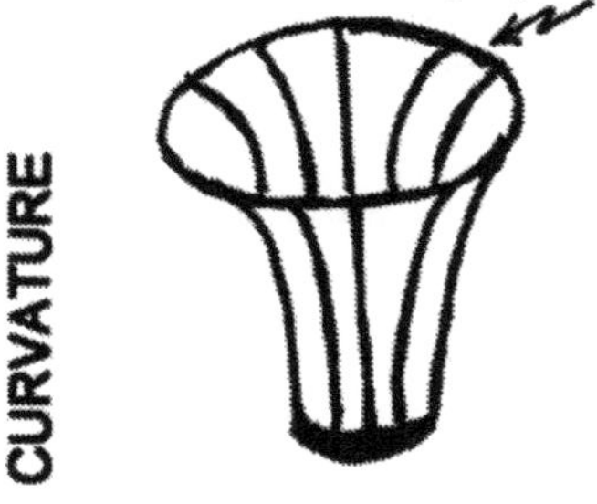

EINSTEIN'S idea of spacewarp as: *lacks motion*. In order to establish "anything that can *exist* in space requires the motion to form it (distance). The electron particle is a form of spacewarp. The endeavor of the sun circling solar system *to square the circle*, causes the square to warp in to a rhombus, which falls in between the shape of a circle and a square, as displayed in the *cird* drawing page 9.

More detailed data is revealed in earlier article named "*PRICE PRINCIPLES OF THE HEISINBERG ZONE*". Some of the subjects in that article concern the #33 on the above-shown electron chart and the 66, as where the 1/3 charge and the 2/3 charge form the electron. Also the quarkery details of most subatomic particles---friction of – static---armatures etc. and tunneling.

The important feature of an electron is that "every" midpoint, between each moving position, possesses the amount 49.5 *increasing* coming and *decreases* going along the configuration as: (49.5 (1 X49.5) 099 (2X49.5) 148.5 3X 49.5) 198 (4X49.5) 247. (5X49.5) etc. This is necessary to account for the stable proton particle shown on page 11 ---- done in circles.

Hydrogens proton is located at the midpoint of the crossover of the + and – current configuration. The description of the proton and neutron is explained after the following display of a free electron.

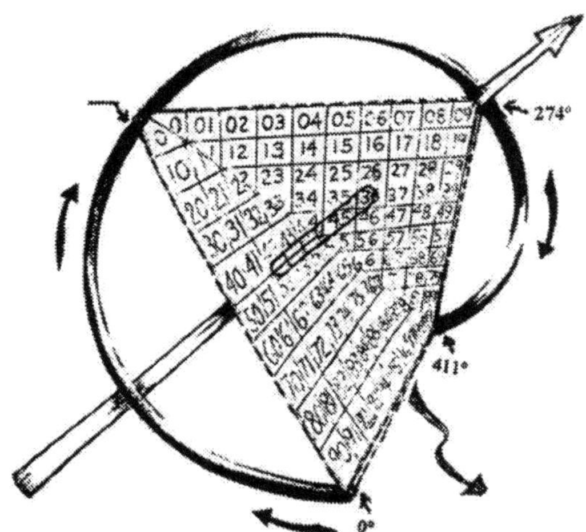

A STILLED MOTION ILLUSTRATION
OF A "CHARGING" FREE ELECTRON
See bottom chart on page 4.

The preceding text went to great length to define the aspects of electrons bound to protons and also free electrons as like the ones flowing in electric current etc. The next paragrahs illustrate the math scheme and geometric display of the momentum/positional change occuring between mesons and protons and neutrons. A proton and its electron co-partner are both stable particles of mass. The nucleus of a hydrogen atom is located at the central midpoint area of the positive field of the atom and contains one proton. Free protons are *rare* and immediately form a current connection that reaches between the positive+ proton and a negative- electron.

As previously explained, the + and – field area surrounding the centered proton, houses a electric current configuration that differs between each position up and down the diagonals by 49.5. As the current proceeds up the diagonal, then back across the top of the 10X10 field, then down through the other diagonal and returns back across the bottom of the field, it feeds a binding quark force to the centrally lcoated proton. The strong binding force shall be covered later in this text.

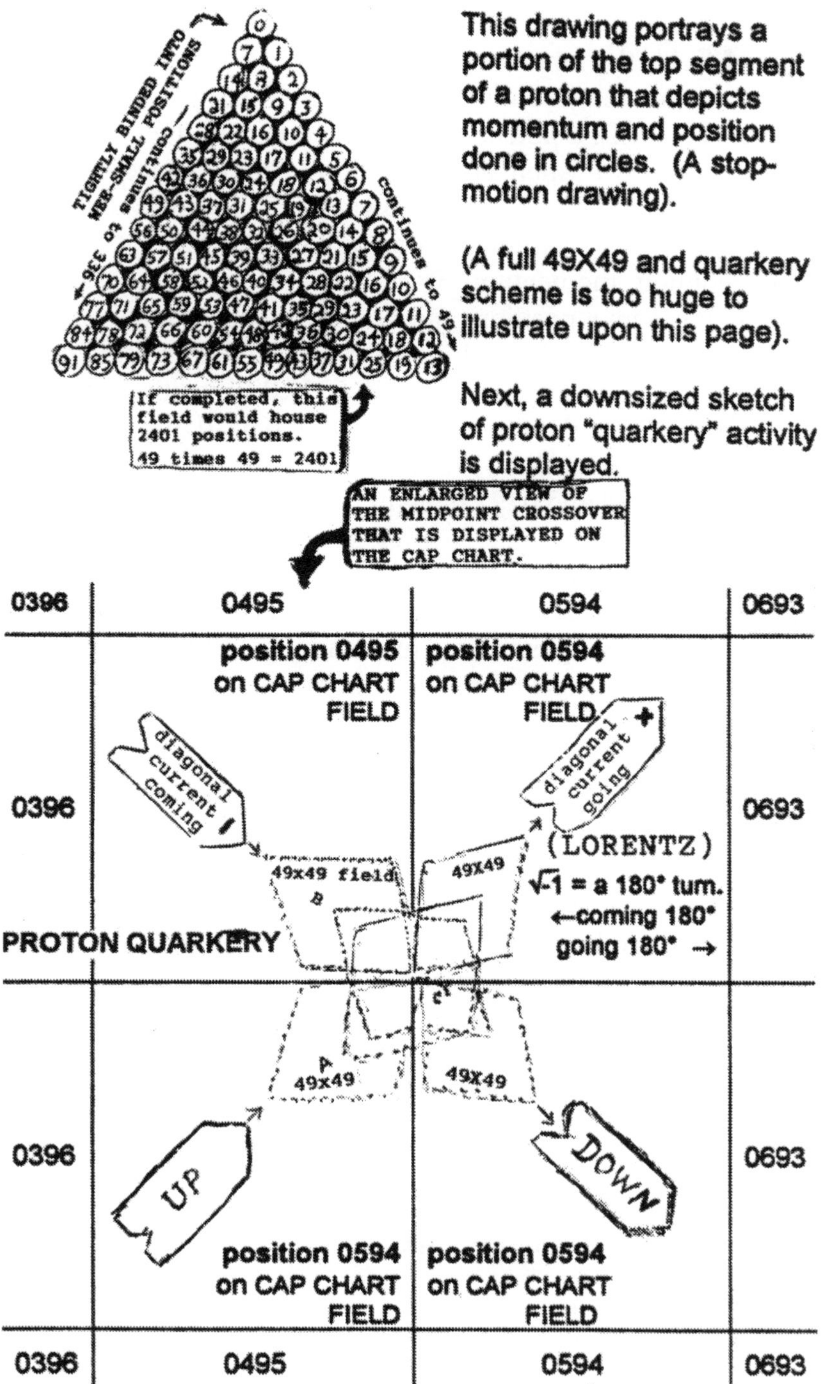
This drawing portrays a portion of the top segment of a proton that depicts momentum and position done in circles. (A stop-motion drawing).
(A full 49X49 and quarkery scheme is too huge to illustrate upon this page).
Next, a downsized sketch of proton "quarkery" activity is displayed.
TIGHTLY BINDED INTO WEE-SMALL POSITIONS
continues to 336
continues to 49
If completed, this field would house 2401 positions. 49 times 49 = 2401
AN ENLARGED VIEW OF THE MIDPOINT CROSSOVER THAT IS DISPLAYED ON THE CAP CHART.
0396
0495
0594
0693
position 0495 on CAP CHART FIELD
position 0594 on CAP CHART FIELD
diagonal current coming
diagonal current going
(LORENTZ)
√-1 = a 180° turn.
←coming 180°
going 180° →
49x49 field
49X49
PROTON QUARKERY
UP
DOWN
position 0594 on CAP CHART FIELD
position 0594 on CAP CHART FIELD

Being that the CAP CHART represents the potential Hilbert orthogonal field, the amounts seen on the CAP CHART are smaller than the amounts that are displayed above in the greatly accelerated scheme. EXAMPLE: a potential position shown on the CAP CHART as 4 becomes 0396 when accelerated by 99, likewise a potential position on the CAP CHART shown as 5 becomes 0495 when accelerated by the maximum amount 99. NOTE, that 0396 also indicates coming momentum of 03 and 96 is going momentum located in the same position and 03+96=99. Likewise 04+95=99. 05+95=99 etc. 99 fills 100 positions due to 0 having the first position. Later a chart that arranges the CAP CHART positions vertically, starting with *0099, 0198, 0297, 0396, 0495, 0594* etc. will be displayed that portrays the 0198 as the 01 "coming" downward and the 98 going upward and 0297 likewise.

0396

↓↑ The 49X49 fields, shown in the above charting, are seen coming through the center in opposite directions, a coming into current and an outgoing current. This action makes the strong nuclear force 137 times stronger then electromagnetic force. Furthermore, when the 10X10 field of an electron reaches the capacity to where it spans out into an electrically arranged scheme of 100 positional rows set into a 100X100 helix field it becomes the starting field that rotates DNA. (Shown later).

The following two diagrams (of 49 converging fields), represent the combined function of that is depicted in the illustration on the previous page of the proton particles quark activity.

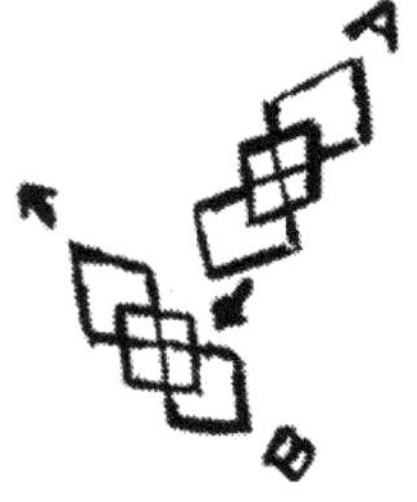

AS A. & B. CONVERGE AT THE MIDPOINT, EACH QUARK SYSTEM EJECTS AN ELECTRIC CURRENT OF 99 POSITIONS. SEE EXPLANATION BELOW. (49X49 CHART)

The area that is where the potential place that can house a proton, is located at the spot that is centered between the positions 05 and 06 on the hydrogen atom segment on the CAP chart. When the potential

(Hilbert space) field positions become accelerated to their maximum energy of 99, the field is a hydrogen atom.

The position 05X99 becomes 0495 and the next position up the diagonal of the chart 06X99 becomes 0594. 0594 less 0495 equals 99. (99/2=49.5). An isotropic system of whole numbered amounts cannot house the addition of .5 into the symmetric system. Therefore, the .5 is ejected from the proton field as:

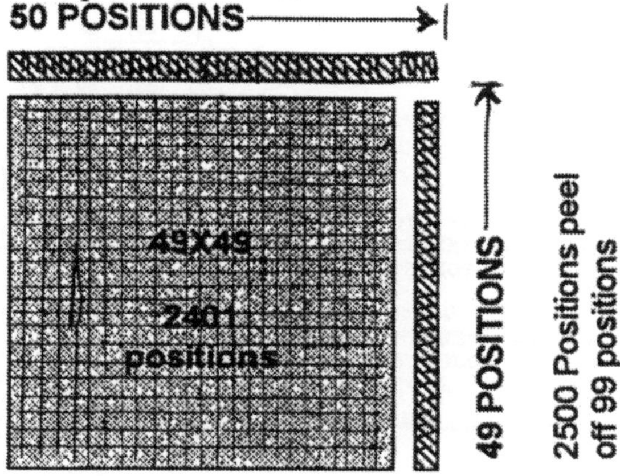

This is drawn in the form of a square. It depicts a thrombus like *form, as is shown in the above* 49^2 *graph.*

The A. & B. combining 49.5 fields that converge and compact at the midpoint, in the hydrogen atom field, gives rise to a larger 50^2 field system containing 2500 position. This larger system exceeds the amount that the field can absorb, and it separates (ejects) 99 positions. 50 off of the top and 49 off the side. The ejected 99, positional current, reverts back up the left diagonal, and proceeds into the far out space where the current flows forms into a negatively charged electron. (See cird diagram displayed on page 9.)

The *results* of the collisions performed in high-energy accelerators and also simple bubble chambers show the subparts of the stable proton. When a proton and anti-proton become stilled in a bubble chamber, the rest mass converts into five pion particles happening as:

$$\bar{P}+P \rightarrow \pi^{+} + \pi^{+} + \pi^{-} + \pi^{-} + \pi^{0}$$

Pi meson exchange scheme displayed in close-packed circles

Pi meson exchange scheme displayed in gridlocked graphed squares

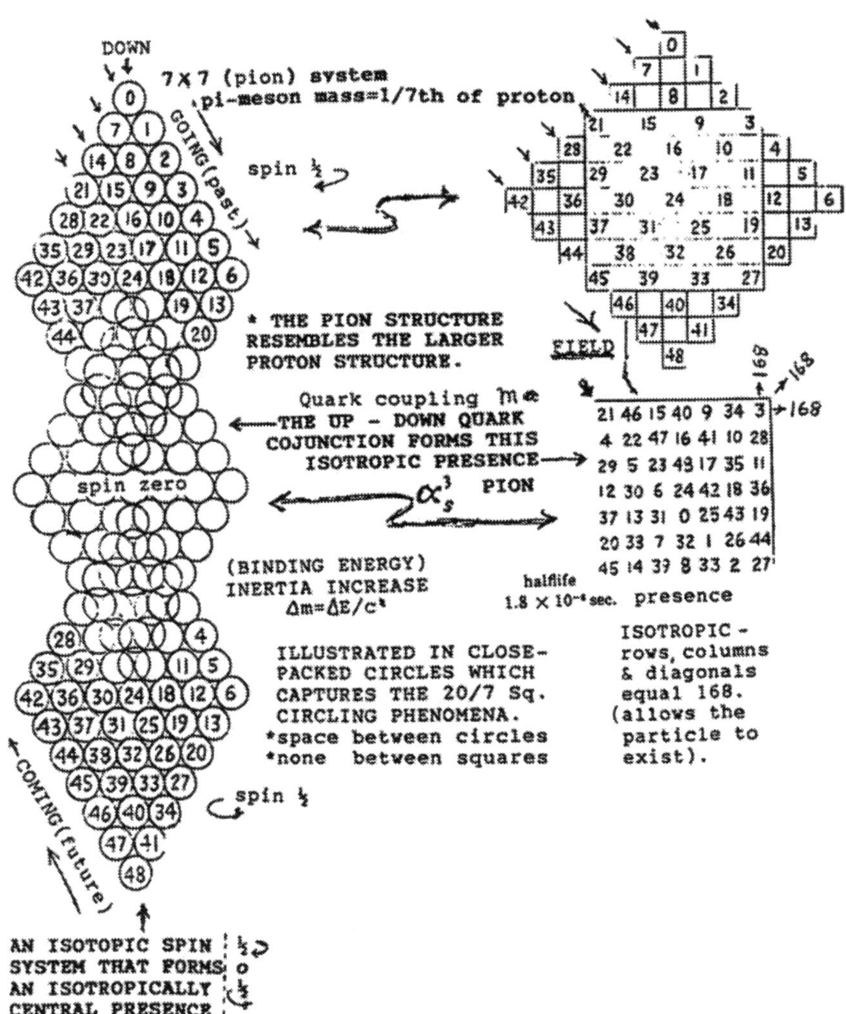

PHYSICIST Yukawa predicted that pions would interact strongly with nucleons. This exchange action is illustrated after viewing the pion on page 14. THE PION EXCHANGE SCHEME IS FORMED BY THE CONJUNCTION OF OPPOSITELY DIRECTED QUARK FIELDS THAT MESH TOGTHER THROUGH A COMMON CENTER WHERE A PRESENCE OF MASS IS CREATED.

A detailed explanation of a few of the many exchange processes will be described after the next paragraph that show the zero-point of electron fields.

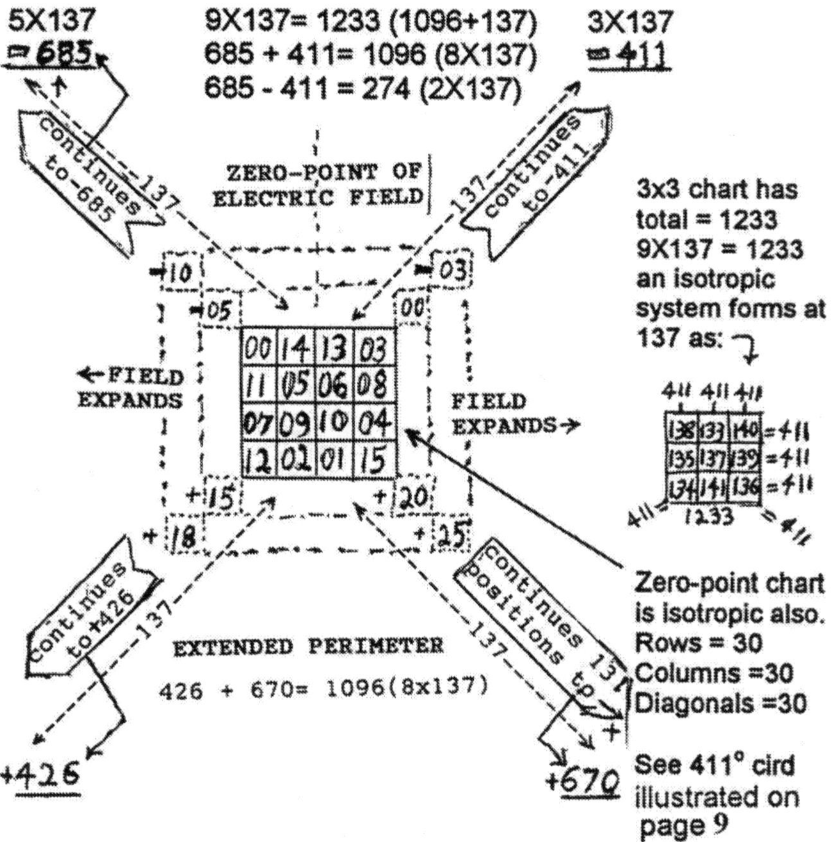

∝ The symbol ∝ refers to the term "fine structure", which determines the close spaced groups of lines that are seen when observing the spectra of hydrogen and helium. The displacement constant is held at exactly or *nearly* 1/137.

The Zero-point graph, that is displayed above, depicts a layout of the field expansion (or decreasing), causing field fluctuations. A field fluctuation is a phenomena known as the Lamb shift. Notice that the diagonals on the 4X4 zero-point cross "through chart" like metric *crossing* decimal as:

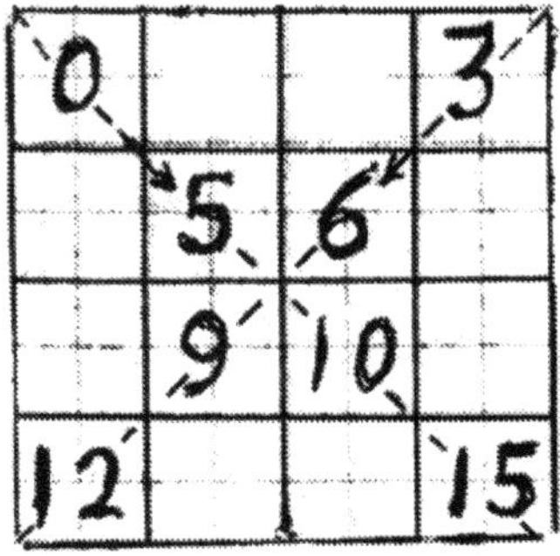

This is an initial function of space-mass development that also ties into the atomic nucleus.

The strong nuclear force is 137 times stronger than the electromagnetic force $(a = 2\pi e^2/hc) \propto = 1/137$.

1941 – University of Princeton – Three professors named Albert Einstein, Peter Bergmann and Paul Bargmann were working on the theory that the electron was a zero point particle formation. They also were attempting to fit the theory of the fourth dimension to apply to a theory by T. Kazula that employed 5th dimensional tensor calculus.

In 1967 I contacted Peter Bergmann for some information that I needed to fit into my zero-point isotropic 4X4 chart. He had written two books about gravitation etc. Bergmann had taken the position of Professor of Relativity at the University of Syracuse in New York. I lived in Pennsylvania.

After conferring by phone and letters, we made arrangements to meet at his classroom one morning at 8:00 o'clock. We become excited that my fundamental math schemes fitted into ideas that the three professors at Princeton had about the point particle scheme of an electron. He went out into the hall and dismissed his class and we worked the blackboard etc. until afternoon. He correctly figured that my point system could only account for the negative force and remarked that it would be ejected→ from the atom nucleus, at light speed, so we quit working on it. At that time, neither one of us realized that "ejection" was what is needed to account for the co-partnership of the proton and *far out* electron. Later in the year, I was walking past a mirror in my lab to throw the 15 chart into the waste basket when I noticed that the 15 looked like a 51 in the mirror. I sat down and worked out the positive (+) system and the midway system that overcame the Heisenberg uncertainty fiasco ($\Delta X \cong \Delta Y$). The *result* gave the math scheme necessary to arrange the atoms as listed on the PERIODIC TABLE in order, along the diagonal of the 100X100 CAP CHART as: (reflection invariance symmetry)

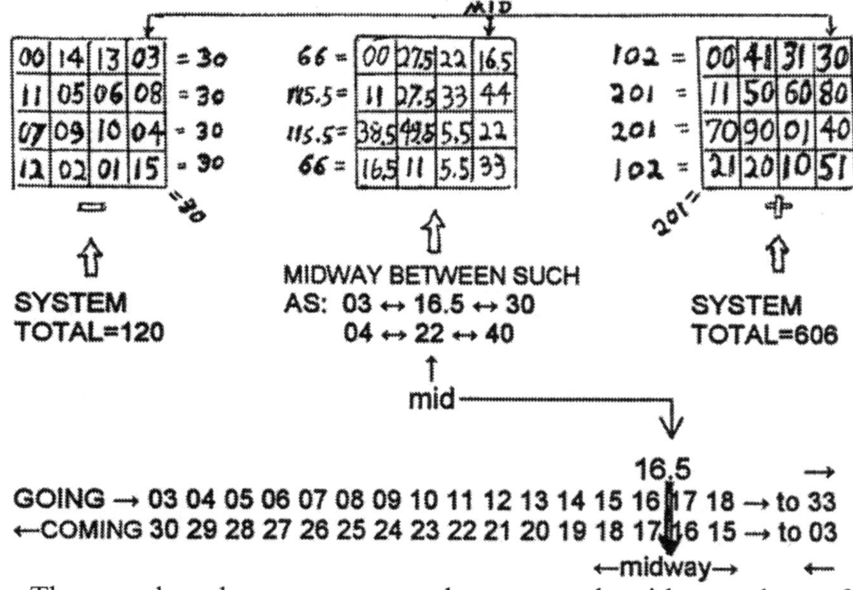

The numbered amounts, on the centered midway chart, fall exactly upon the nucleus crossover points that designate the different atoms that form up the diagonal of the CAP chart. This coincides with the periodic table of elements. Chart numbers ---5.5 (H)11 (He) 1.5 (Li) 22 (Be) 27.5 (B) 33 (C) 38.5 (N) 44 (0) 49.5 (F) SEE CAP CHART.

THE NUCLEAR FORCE EXCHANGE MECHANISMS

Review the pion exchange particle, done in circles, displayed on page 11. Notice that the converging quark fields are spinning in opposite directions! Coming/going. DOUGS LAW – At the weesmall realm in the Heisenberg zone, where things come together into being, any consecutively arranged field or a squared segment of the field, can be formed into an isotropic system structured relative to the manner that it is observed.

THEREFORE, the spins of the converging pion quarks are able to reform the math scheme of the field that it overcomes the Heisenberg uncertainty fiasco, $\Delta X \cong \Delta Y$, which also holds for the difference of dimension and position. $\Delta D \cong \Delta P$. Examples:

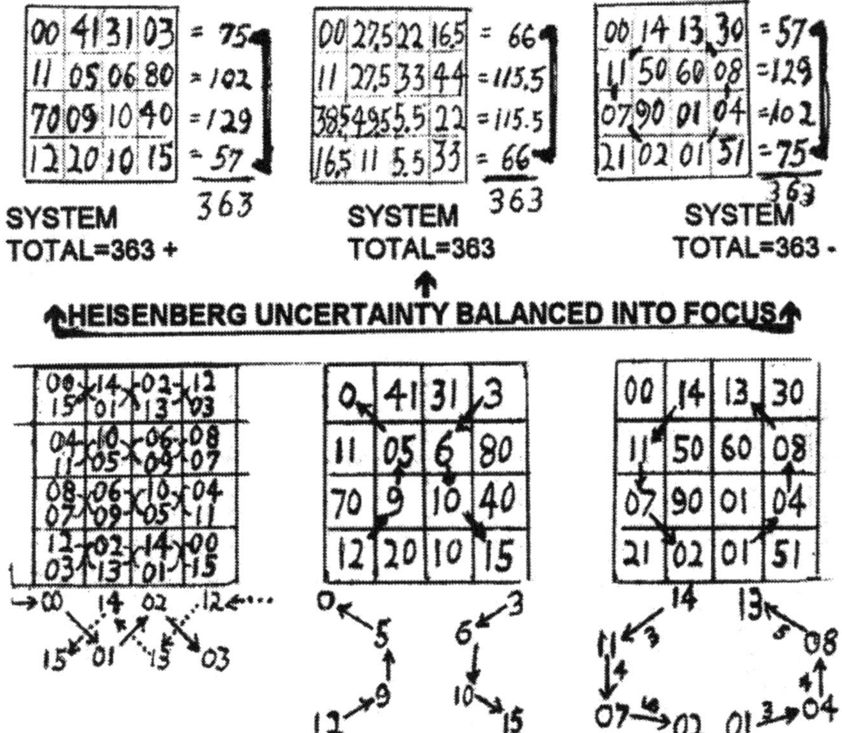

↑This chart depicts a 2-D scheme of the jointure of two converging fields – 1 *coming* 1 *going*. This field action makes up the zero-point chart shown at top of pg. 19

↑↗These two charts depict the result of two converging quarks having opposite spin. Note that the energy is consecutive. 345 going → and ← 543 coming. Study reflection invariance and conversation of parity.

ESTABLISHING A PRESENCE

COMING/while/GOING

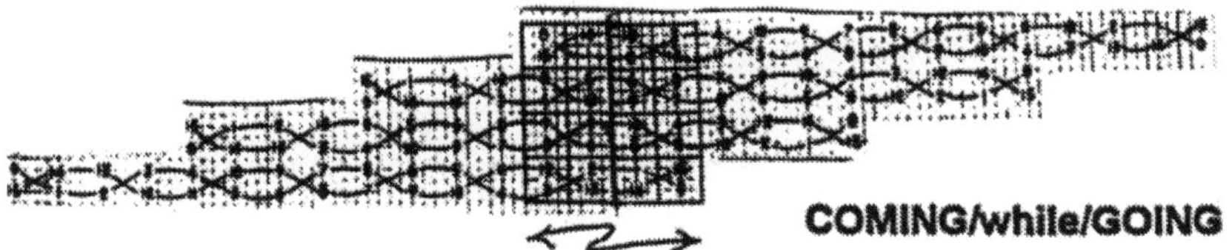

Zero-point base of static electricity, like was called, "The Blue Ghost", by sailors referring to the blue flame-like waves criss crossing the vessels sails.

Shown above as reciprocating to right (then it shifts to left).

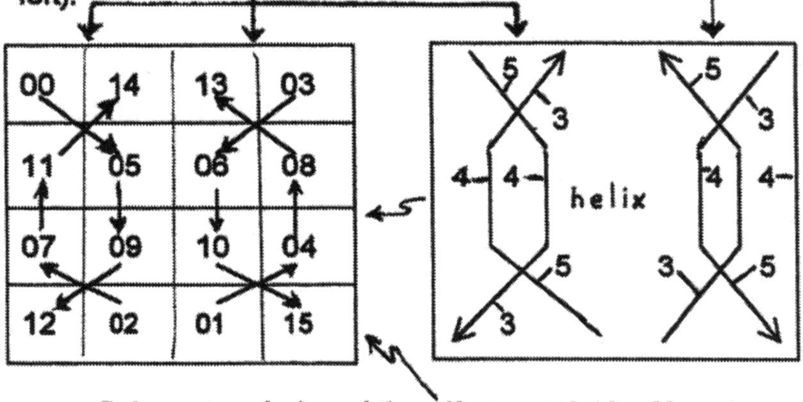

00
14
13
03
11
05
06
08
07
09
10
04
12
02
01
15
5
3
4
4
helix
5
3
5
3
4
4
3
5

Spin energy induced thru 4^2 zero-point by 3^2 system.
ARISTOTLES CONSECUTIVE POSTULATE

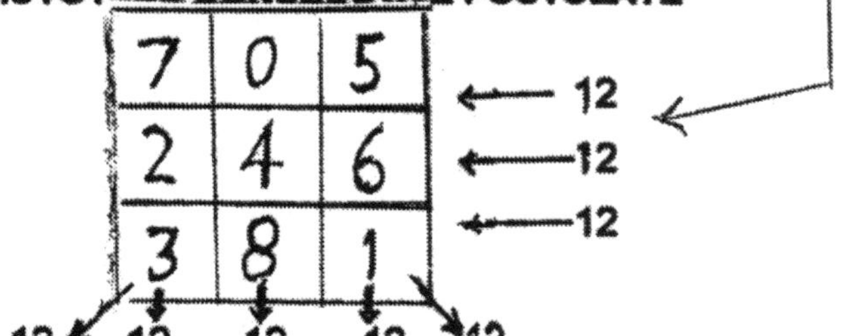

7 0 5
2 4 6
3 8 1
12
12
12
12 12 12 12 12

3^2 system is a symmetric tensor of the 9th rank having 81 vector scalar compents

Max
Plancks
h
↘

						4	0	8
						2	7	3
						6	5	1
			7	0	5			
			2	4	6			
			3	8	1			
4	6	2						
8	1	3						
0	5	7						

Energy as a 3X3 system, going into mass as the 4X4 system requires much explanation [See *Price Principles of the Heisenberg Zone* text).

Next – The exchange of nuclear forces at the pion range.

The pion drawing (done in circles) page 14, depicts the charting of *momentum* changing *position* in coverging quark fields. This converging action forms a "presences" of mass that exist having a half-life of 1.8 X 10^{-8} seconds.

The chart illustrated next done in squares, represents the isotropic *product* of the converged quark fields (presence). THE DECAY MODE $\underline{\pi} = \underline{e^{+} + v}$ is displayed on page 21.

168

21	46	15	40	9	34	3	←
4	22	47	16	41	10	28	←
29	5	23	48	17	35	11	←
12	30	6	24	42	18	36	←
37	13	31	0	25	43	19	←
20	38	7	32	1	26	44	←
45	14	39	8	33	2	27	←

168

This *totally* isotropical arrangement is what makes the pion able to exist for a brief time. PRESENCE

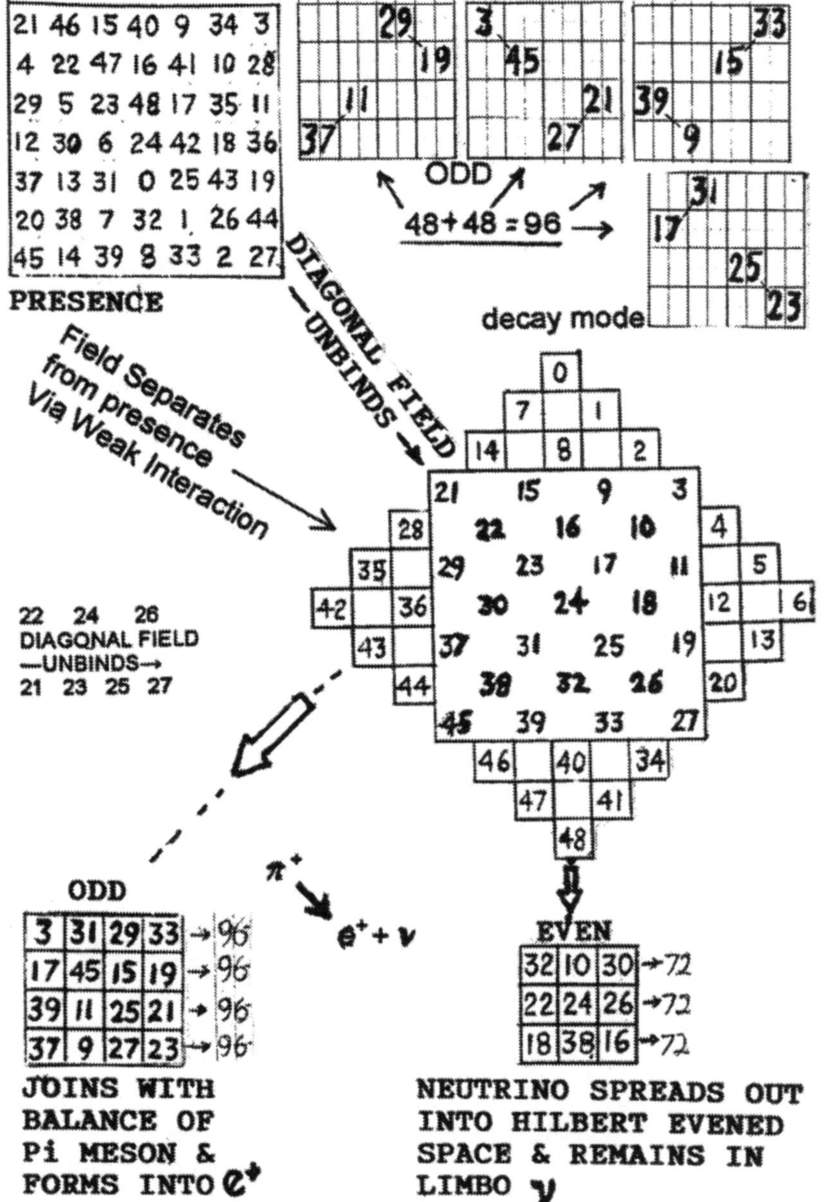

The muon mass is 1/8 that of a proton. –pion is 1/7 th. The above-pictured charting, of the decay mode of the pi meson particle, signifies that the neutrino (3X3 chart) sets off *apart* from the meson, and becomes "stilled" in the Hilbert orthogonal space surround. π^+HAS HALFLIFE OF 1.8 X 10^{-8} sec. In order for a particle to be "detected", the system must be structured with both even *and* odd amounts in its makeup. Therefore, neutrinos can be found anywhere and everywhere in Hilbert space, but cannot be detected --- unless they are activated by *odd* amounts added to their system. When this "*odd*" activation occurs, then they are not neutrinos as such. When the 4X4 chart "sets off" apart from the meson (called "emitted" nowadays), the balance of the meson ties in with space causing it to become a positron $^{+}$ The position is established by the isotropically arranged 4X4 chart shown above. Notice that the 3X3 system contains all *even* amounts of momentum. The sequence of events happen as $\pi^{+} \rightarrow e^{+}$ or $\pi^{-} \rightarrow e^{-}$ + The π - meson involves 147 positions (3X49). See the pion drawing (displayed on page 14 in circles).

The decay mode of the meson requires additional positions to form a *positron*$^{+}$from a 4X4 charting. The position requires 198 positions (99 coming --- 99 going).

THE 3-DIMENSIONAL CARTESIAN COORDINATE X-Y-Z SYSTEM IMPLIES THAT SPACE IS END LESS. EVENLY AVERAGED OUT. NEUTRINOS SPLIT ODD FROM EVEN & BECOME STILLED AS:

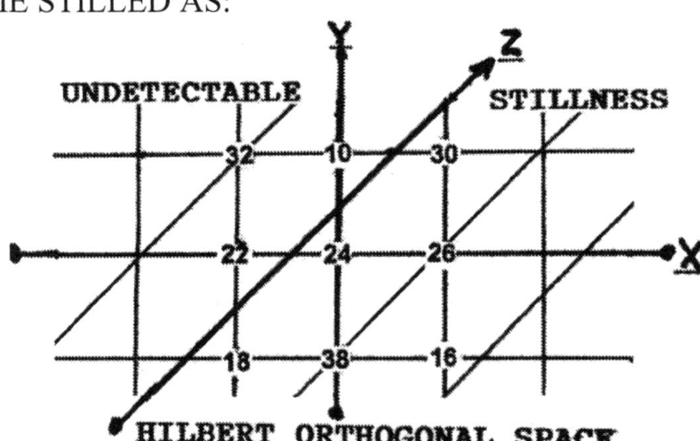

PARTICLE DECAY MODES – [The relativity connection]

The Heisenberg Uncertainty Principle ($\Delta X \cong \Delta Y$) is overcome by Karl Gauss type math that crosses a common center.

	0 1 2 3 4 5 6 7 8	(going) →
← (coming	8 7 6 5 4 3 2 1 0	

↑

The above-shown math scheme depicts a midpoint balance at 4. Therefore, momentum and position *can* be balanced in a system. A decay process of a sub-atomic particle *must* occur in a manner that conserves momentum. In order for this to happen, the particle itself, and its emitted counterparts, must have the momenta in its system isotropically arranged. Therefore, a particle is able to "*exist*" because its isotropic arrangement of momentum is the same, relative to the viewer, probed from any direction. Isotropic arrangement allows a system to maintain a brief presence. Space, as a position, is *apprehensive* to a "probe" of momentum, → (relativity). The pi-meson and its emitted particles are totally isotropic in momentum arrangement. Charted as:

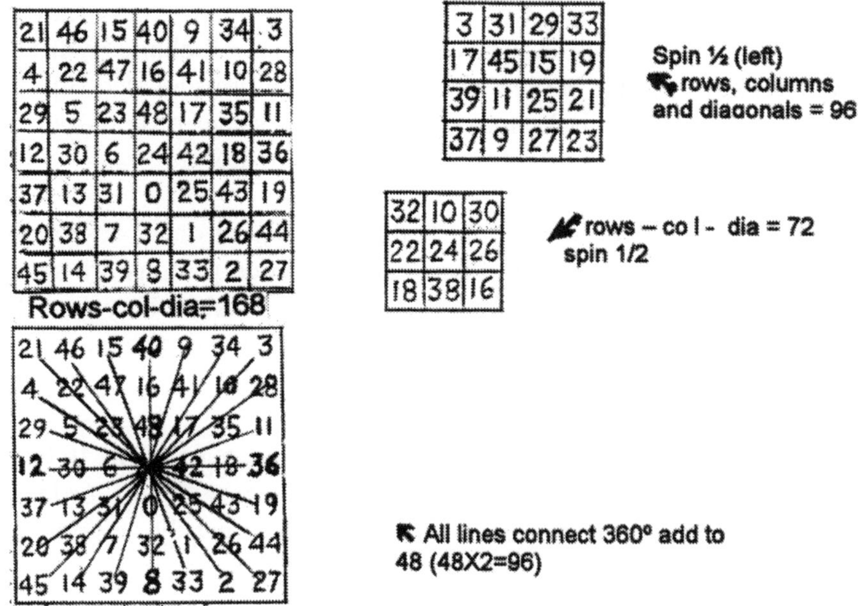

Another conservation factor, concerning decay modes, is the control of the amount of momentum in the diagonals from the common center of the system. → The virtual momentum is conserved as:

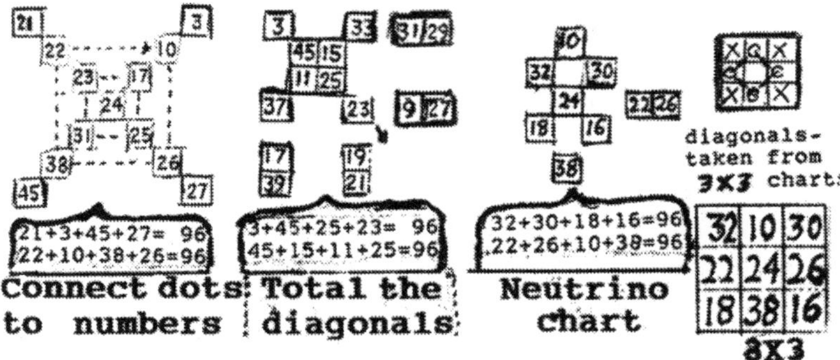

THE *DIAGONAL PERIMETER* REMAINS CONSTANT THROUGHOUT THE ENTIRE PI-MESON DECAY PROCESS MAINTAINING "CONSERVATION OF MOMENTUM".

Review the 360 degrees chart (above) and note that the pi-meson is completely isotropic as 12+36+40+8=96 and 6+42+48+0=96 etc. no matter *where* you view the system. The hydrogen atom is housed in 100(10X10) positions shown on the potential CAP chart. Hydrogen contains *one* proton. Notice that the next atom, located up the diagonal of the chart, is helium. Helium houses *two* protons but, its *size* is nearly four times larger than the hydrogen atom!

Like forces repel (+←→+). Unlike forces attract (+ →← -). Therefore, there is a problem when two like force protons are seated together in the nucleus of an atom. Thus, there must be a system that neutralizes the like forces of protons and permits the dual existence of protons within the nucleus.

The neutralizing system is a particle named a neutron. It has a half-life of about twelve minutes. $N \rightarrow P + e^- + \bar{\nu}$. In order to "neutralize" an energetic force, the neutron *must* be completely versatile and totally isotropic throughout.

A portion of a neutron exchanges with a portion of a proton. The manner of exchange is via of the pion particle and a system (charted by Ben Franklin) that contains 256 positions (16X16). The pion particle contains 49 positions. The Franklin charted system is a *segment* of a large neutron. The Franklin chart is also totally isotropic, and is the pion. The virtual momenta is exchanged and neutralized as:

FRANKLIN CHART

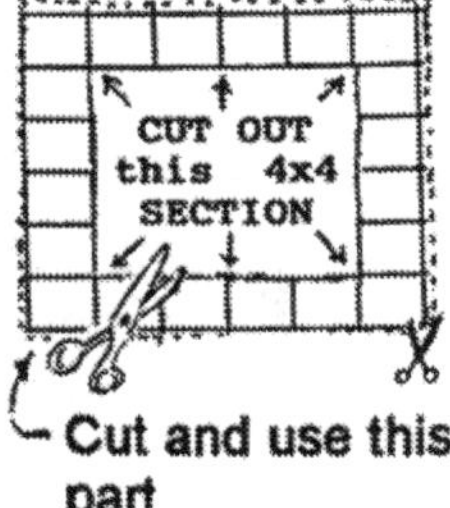

199	216	231	248	7	24	39	56	71	88	103	120	135	152	167	194
57	38	25	6	249	230	217	198	185	166	153	134	121	102	89	70
197	218	229	250	5	26	37	58	69	90	101	122	133	154	165	186
59	36	27	4	251	228	219	196	187	164	155	132	123	100	91	68
200	215	232	247	8	23	40	55	72	87	104	119	136	51	168	183
54	41	22	9	246	233	214	201	182	169	150	137	118	105	86	73
202	213	234	245	10	21	42	53	74	85	106	117	138	149	170	181
52	43	20	11	244	235	212	203	180	171	148	139	116	107	84	75
204	211	236	243	12	19	44	51	76	83	108	115	140	147	172	179
50	45	18	13	242	237	210	205	178	173	146	141	114	109	82	77
206	209	238	241	14	17	46	49	78	81	110	113	142	145	174	177
48	47	16	15	240	239	208	207	176	175	144	143	112	111	80	79
195	220	227	252	3	28	35	60	67	92	99	124	121	156	163	188
61	34	29	2	253	226	221	194	189	162	157	130	125	98	93	66
193	222	225	254	1	30	33	62	65	94	97	126	129	158	161	190
63	32	31	0	255	224	223	192	191	160	159	128	127	96	95	64

x

The current weaving back and forth across the Franklin 16^2 chart moves consecutively. Notice at 160, positioned on the bottom row, weaves to 161 then back to 162 then to 163.

Cut out the blank section from the graph – SLIDE IT any *anywhere* upon the Franklin chart, and the total showing thru the cut open section will equal 2040 (15X137=2055) pion exchange adds 15 to the 2040

The Franklin segment of a neutron chart, is capable of exchaning force relative to *any* directional observation mode. When the neutron is out of its atomic field area, its decay mode last about 12 minutes. The momenta timing sequence interweaves throughout this "flat" charted graph form staring at the base of the chart, and proceeding upward as:

1 3 After 3, look upward, 3/4 of way to top of chart → 10 8

0 2 "going" up and see "coming" down ↓ 11 9

The Franklin chart weaves its current course this way from the 0 starting pt. and ending at 255 (beside 0 at bottom).

Total of each row is – 2040 (15X137=2055) pion exchange adds 15
Total of each column – 2040 Total of each diagonal is 2040
Total of any 4X4 group *anywhere* on the chart is *2040.*

2040 This permits numerous ways for the manner of exchanging and neutralizing the like forces of protons in the nucleus.

The pi-meson, called a pion, is the system that functions within the nucleus of an atom. It manipulates the exchange of the strong forces that are possessed by protons which contain an outward (+positive) force. Like forces repel each other strongly.

The pi-meson is well named because it incorporates the irrational circle-squaring fiasco (π) in its force exchaning maneuvers. The pion illustration, done in circles displayed on pg. 14 could also have been shown in squares. Also it could be done in rhombus forms, which are like ½ way between a circle and a square (probably the best way to perceive curve fitting). The amount conceived as π is natures method of endeavoring t. balance (focus) the difference between a circle and a square. The irrational math scheme of π is 3.141592 --------- *endless* taken to its nearest fraction is 22/7 for circle-squaring and 20/7 for square-circling. How does this circle-squaring endeavor appear geometrically?

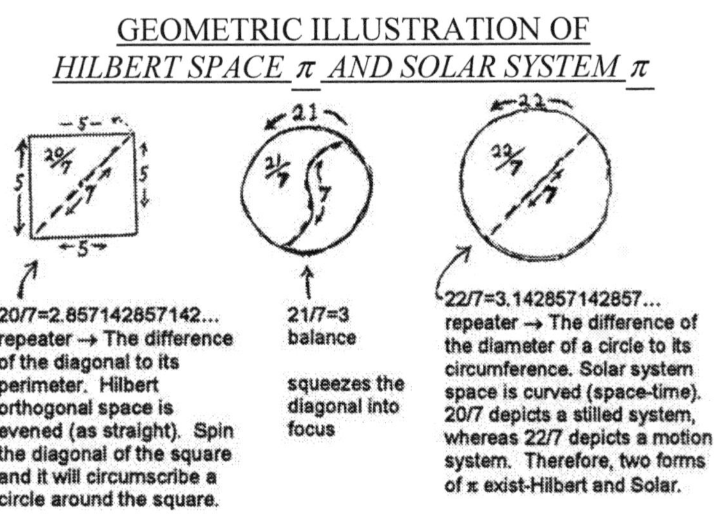

The pion is 1/7 the mass of a proton. A neutron is slightly larger than a proton on a ration of 8 to 7 (mass). *NEUTRON*

The neutron occupies 4096 positions in space. The chart, that is displayed below, is taken from the top left corner of the large 64X64 chart that is illustrated in the earlier Book called ***The Price Principles of the Heisenberg Zone***. The momentum in the neutron changes position in isotropic manner.

A PARTIAL GRAPH OF CURRENT DISTRIBUTION IN A NEUTRON PARTICLE

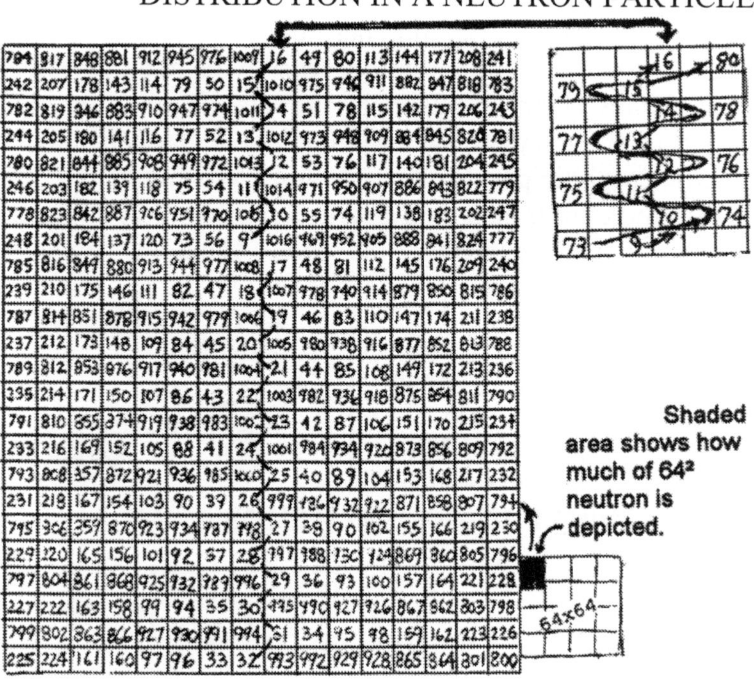

The complete 64^1 neutron chart is too large to be displayed in this book. Study this *top portion* of the 64^2 chart to be familiar with the locus of the current flow.

The above chart shows a portion of the current that is moving all throughout the 64X64 neutron chart. Look nine positions across the top row of neutron chart and trace the flow of the current schem

e.

Use the template cutout from the FRANKLIN chart. Slide it anywhere across the partially displayed neutron graph and find that the amount that shows through the aperture is 8200. The rows-columns and diagonals of the entire chart total 8200. The strong nuclear exchange force can employ many pathways. The strong nuclear force is 137 times more stronger than the electromagnetic force.

See the Franklin chart displayed on page 25.

Note that each row on the Franklin chart, carries the total of 2040 (15X137=2055). Also, more importantly, when you slide the 4X4 positional template anywhere on the chart, the total showing through the cut-open part of the template will be 2055. This can account for the neutron particles participation in the action exchanging the 137 strong force. But, there *must* also be a 137 group endowed within the pion particle in order to exchange the force. This exchange is illustrated as:

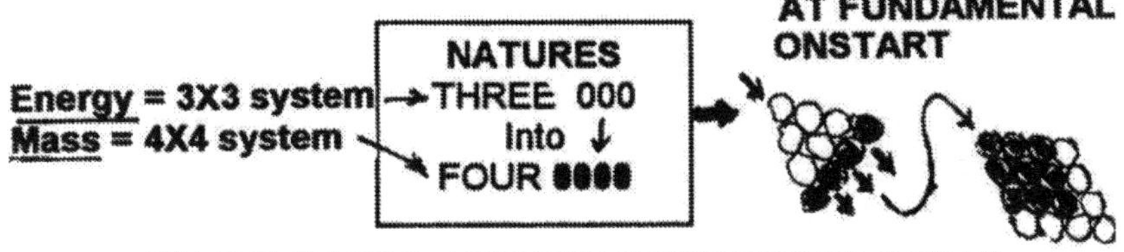

ENERGY THING INTO THE 7^2 SYSTEM OF A PION

This exchange resume shall begin with the exchange process that occurs in the helium atom because hydrogen has no neutron in its nucleus. Hydrogen will be discussed later.

195	220	227	252	=	894
61	34	29	2	=	126
193	222	225	251	=	894
63	32	31	0	=	126
					2040

A segment of the Franklin chart displayed on page 25

All 4X4 segments total to 2040.

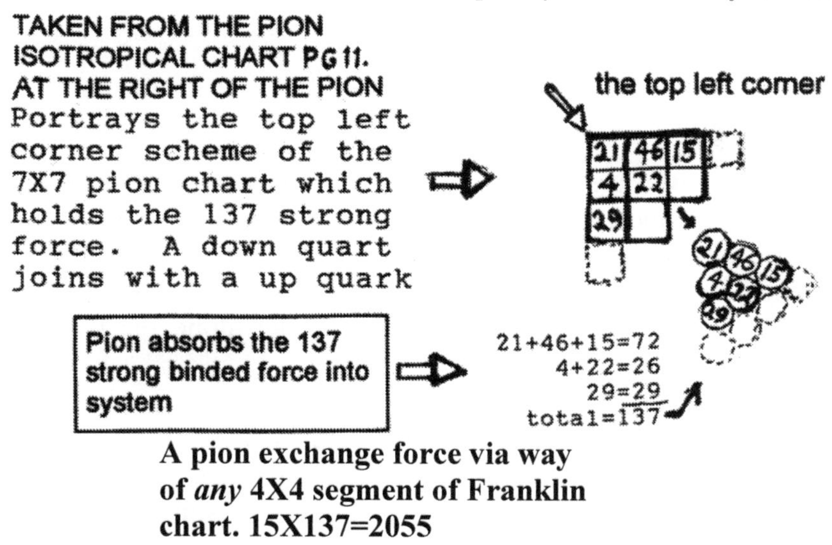

A pion exchange force via way of *any* 4X4 segment of Franklin chart. 15X137=2055
2040 + <u>zero – pt.</u> (15) = 2055

Helium possesses two protons and two neutrons and two (far out) orbiting electrons. See CAP chart and note that helium is about four times larger than hydrogen. Displayed above, are some of the fundamental systems that enter into the strong (137 X electromagnetic) nuclear force exchange process.

A hydrogen atom is housed in 100 positions (see CAP chart). In order for the zero-point electric field to be able to exchange the 137 strong force, the neutral field area must total 256, as in the Franklin chart that is displayed on page 25.

Therefore, hydrogen has no neutron.

<u>Zero point of</u>

Electric Field Total = 120 4X30=120	Momenta of 120, surpasses the 100 field. The momenta in the 100 field = 99. Relativity's quantum observation resets the 120 back into the 10 (10X10) mode.

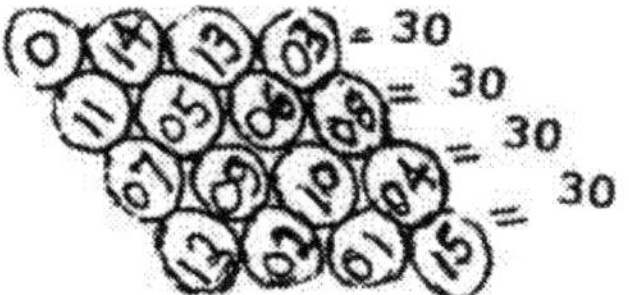

It would be nice if the activity that is happening among the systems in the nucleus of atoms could be explained like a recipe that adds the ingredients to a cake. But, alas, it is a ***jungle*** within the atom. --- Repelling forces-short circuits violation of symmetry – parity --- resetting broken laws of conservation etc., let alone the backwards observation modes of relativeity!

Back in the era of the 1920's, the "Old Masters" were striving to attempt to chart a model of an electron particle, using math of dimensions – proportions and momentum – position to chart the system of an electron, Werner Heisenberg came dancing into the picture, suggesting that, "you old masters cannot accomplish your charting of an electron because there is *uncertainty* within the realm of an electron!" Now what? So, we are left with a choice --- Run away from the atomic jungle like scarred rabbits, or pretend that we are brave, and remain working with the old masters, endeavoring to chart sub-atomic data. (I hope that I'm not the last of the dying breed of the great pretenders).

The charts that follow portray a few of the happenings that occur during the changing of forces in the nucleus of atoms. The larger the atom the more intense is the nuclear activity. The next shown chartings portray short-lived systems that adapt to natures laws of conservation.

At this time, the writer is going to take the reader to a place that you have never been. It is called Nowhereville. Issac Newton, in answer to the question, How does charge A know what charge B is doing, since A and B are apart from each other? Newton's reply was that, "the action at a distance is simultaneous since there is nothing between charges A and B". Einstein rebuked the simultaneous idea with his postulate "Nothing can exceed the speed of light." Who was correct? If the reader goes down to the no-where of space and begins to move at 186,274 miles per second from position 1 toward position 2 and continues upward to position 3 (which is already there), how fast is already there? Simultaneous! The co-junction of Newton's thinking and Einstein's light-speed idea allows for the Law of Conservation of Momentum. Light contains momentum, so, if nothing exceeds the speed of light, then, "nothing" must be position (space). Momentum requires position in order to exist. Position exists with the need of momentum. BUT, in order for position to be established as such, it must be designated by momentum, otherwise position is the Nowheresville in space. So be it. --- A geometric description of momentum done in circles is as:

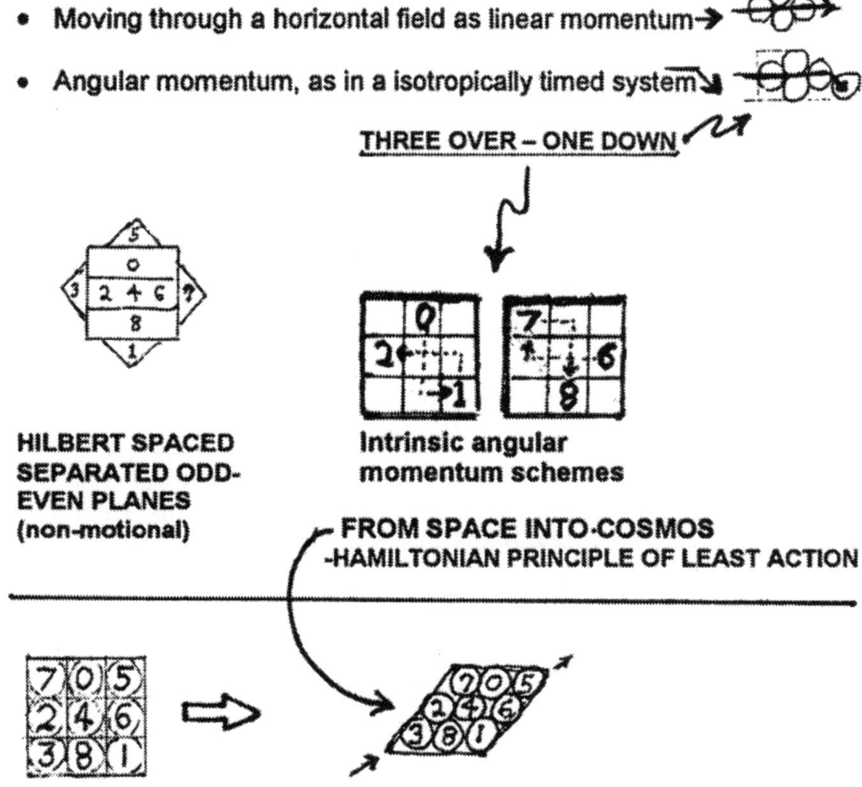

POTENTIAL ACTIVATED BY THE HIGGS MECHANISM (MULTIPLICATION)

•Before proceeding to unravel the exchange of repelling strong forces within the atoms nucleus, it is absolutely necessary that the reader grasp the significance of opposite directed converging fields being able to establish a brief "presence" into becoming an isotropically *timed* atomic particle system that endures an-existence in the cosmos.

We begin, as did Albert Einstein, by accepting the "all evened" Hilbert orthogonal space, as the all scalar background from which mass is derived. Next, odd is mathematically, added to the Hilbert evenness. The addition of odd amounts, like 1-3-5-7 etc. to evened space like 2-4-6-8 etc. allows the Hilbert space scenario to acquire direction (a vector). Space, as such, is no longer Orthogonal. Space, being endowed with direction, becomes capable of field production that sets up the potential background for the placement of atoms into existence. This potential field is what is referred to as a Higgs Field, but not known, as such, until now that I reveal it. The Higgs Field is the exact motionless odd-even background arrangement that is enhanced with both magnitude and direction. A Higgs Field is the spacing *arrangement* that, when accelerated into motion, fills the place in space, where the potential particle is situated, to acquire mass and become a particle of mass.

The CAP chart portrays the Higgs potential arrangement of the periodic table of elements (from hydrogen to neon). FOR INSTANCE: The Higgs mechanism comes into play when the potential field area is accelerated (multiplied) by 99. This acceleration performs an action called the Higgs mechanism, which gives the potential system a "presence of time" that brings the potential particle into existence as an inertial body of mass.

The layout of "presence of time" acquired by the atoms (from the action of the Higgs Mechanism) is illustrated next.

POTENTIAL ARRANGEMENT OF ATOMS IN A HIGGS FIELD AS ON CAP CHART	The Higgs mechanism fills the field positions to its limit by multiplying X 99	HIGGSED RESULT (for hydrogen see Higgs chart on next pg.

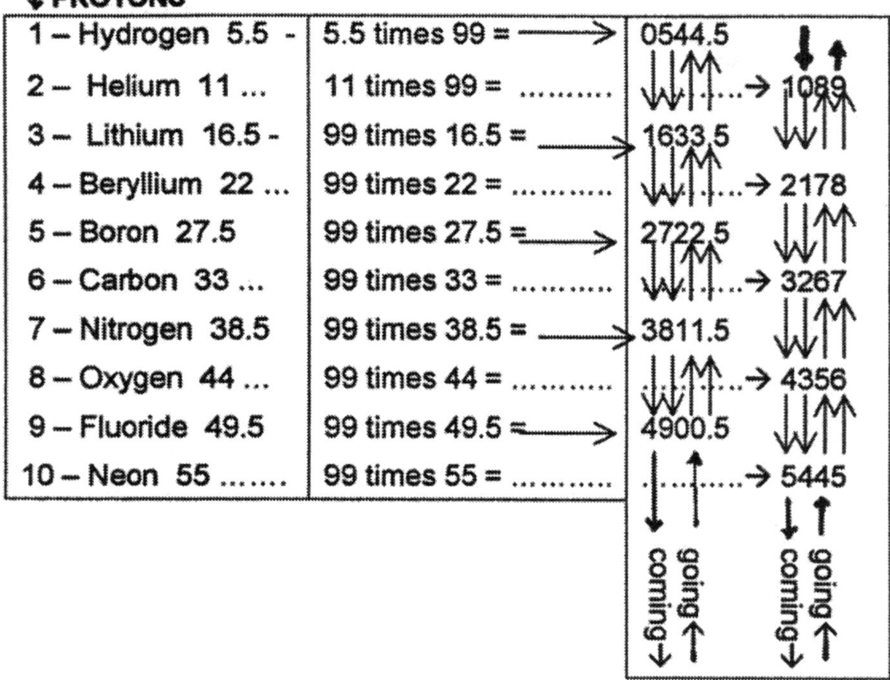

AMOUNT OF PROTONS

1 – Hydrogen 5.5 -	5.5 times 99 = →	0544.5
2 – Helium 11 ...	11 times 99 = ...	...→ 1089
3 – Lithium 16.5 -	99 times 16.5 = →	1633.5
4 – Beryllium 22 ...	99 times 22 = ...	...→ 2178
5 – Boron 27.5	99 times 27.5 = →	2722.5
6 – Carbon 33 ...	99 times 33 = ...	...→ 3267
7 – Nitrogen 38.5	99 times 38.5 = →	3811.5
8 – Oxygen 44 ...	99 times 44 = ...	...→ 4356
9 – Fluoride 49.5	99 times 49.5 = →	4900.5
10 – Neon 55	99 times 55 = ...	.....→ 5445

Note that each atom is separated by 5.5 (proton). The coming while going saga, acquires some past time and some future that forms into a "presence" of time, when the potential atom on the CAP chart becomes activated by the Higgs mechanism. The 100X100 field is filled to capacity, with protons, at neon. Also, notice on the CAP chart, that all of the positive+ charged configurations return back through the 10X10 hydrogen field.

The next chart represents the positive+ charged field area of a hydrogen atom. The atom has been *activated from its Hilbert potential place* in space by the multiplication action of the Higgs mechanism. The chart, shown next, is a segment taken from the huge DNA starter chart, that is illustrated in the "*Price Principles of the Heisenberg Zone*". Notice the way the current is aligned, in a coming/while/going manner, along the top row of the chart. The shape of a DNA molecule is a double helix that has a coming/going/current field crossing back and forth, between the two helically moving strands. The amino acid molecules become activated when the crossing current contacts their specific potential positions in the field area that enacts the Higgs mechanism process. (See Heisenberg Zone text).

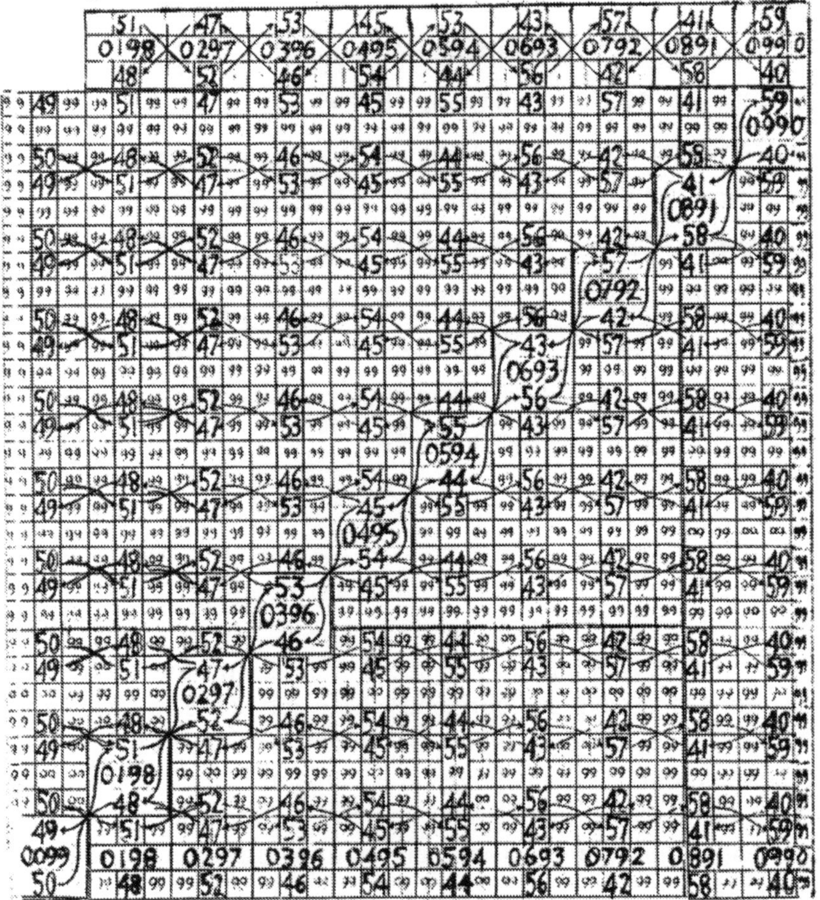

COMPARE THIS CHART WITH THE HYDROGEN *POTENTIAL* CHART THAT IS AT THE BASE OF THE DIAGONAL OF THE CAP CHART. THE CAP CHART IS LIKE A LENGTH OF COPPER WIRE ---WHEN THE COOPER WIRE IS ACTIVATED BY CURRENT, IT BECOMES LIKE THE ABOVE SHOWN CHART.

Notice that the positive charge area is the same math scheme as the diagonal position that is directly below it as:

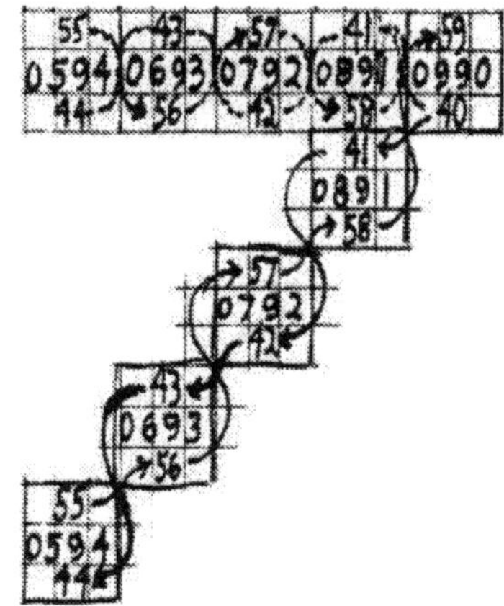

The current running up the diagonal has a longer way to go from bottom to top of the chart than does the *top row charge* that is jam-tighter together making a charge. $\frac{20}{7}$ Also notice that the positions all conserve the 99 momentum that is the limit of the 10X10 (100 positional field area as:

0099	0198	0297	51	47	53	45	
00	00	02					← COMING WHILE
99	98	97	48	52	46	54	GOING →
99	99	99	99	99	99	99	

The following paragraph is written so that the aspects of an electric current can be better understood. Also to acquaint the reader with an explanation of how time is induced into a sub-atomic system making it able maintain an existence in the cosmos.

Back in the 1920's, when the Old Masters were attempting to structure a model of an electron, using momentum changing position and math of dimensions and proportions. The brilliant German physicist, **WERNER HEISENBERG**, came up with the fact that momentum and position is not able to be focused as such. When you examine the position, the momentum has already vacated the position, and vice versa. Heisenberg instilled the quantum theory into our world.

Now, something even complicates the $\Delta X \cong \Delta Y$ matter worse! That is, in order for something to exist is space *ALSO* requires time. Time, added into a place in space, that allows a thing to contain being, must be *isotropically* structured in space having ½ of the system *coming* from the future AND ½ of the system *going* to the past. Simply stated, some of the time must be obtained from the coming future and some of the time must procured from the past.

(Geographically shown below). this was introduced on pg. 4 Also on the FREE ELECTRON CONFIGURATION. pg. 9.

Something happens ½ way up the diagonal of the CAP chart, that *changes* "going into coming" and the "coming into going" in a manner that brings momentum into focus with position.

QUESTION: Why does something happen on the diagonal of the chart.

ANSWER: Lepton particles, such as the electron that is in the 10X10 system displayed on pg. 9 manipulate (coming/while going) *horizontally* across a field.

Protons, pion etc. family, maneuver diagonally in a field, which promotes the binding force. The following diagram portrays the crossover area of the diagonals, where the "coming from" + current line changes into a "going to" – line of current. ½ way midpoint crossover shown below as:

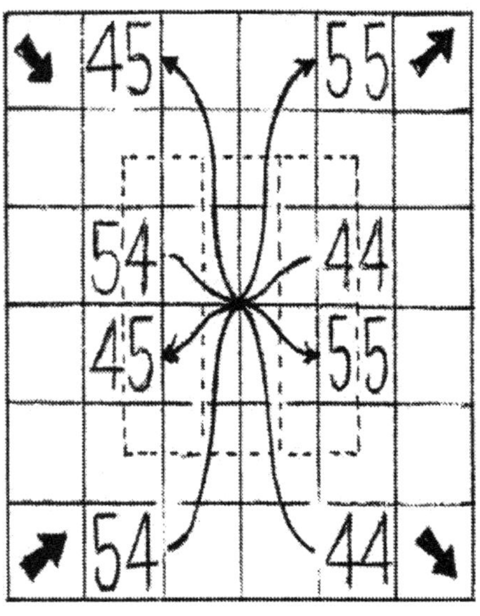

Enlarge drawing, of the central four positions at the crossover, of the diagonal current in the hydrogen atom chart that is located on pg. 34

54→55 45→55 diagonal
54→55 44→55 horizontal

All crossing through the 49.5 midpoint center.

The chart *beneath* this one is also

an enlarged drawing of the horizontal current that crosses back and forth across the field while conserving momentum at

99. 04 94
<u>95</u> <u>05</u>
99 99

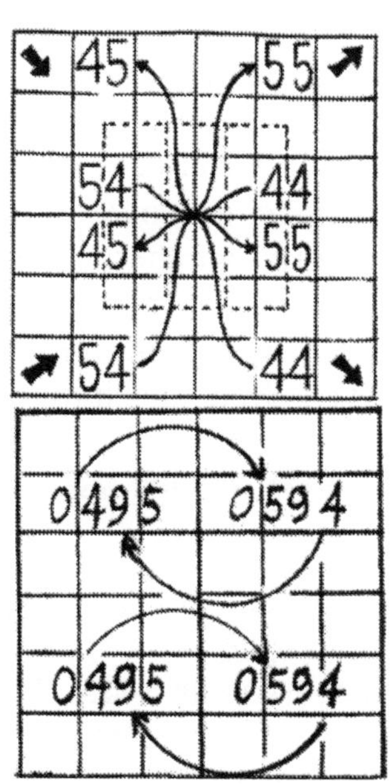

Here again, the quantum observation mode confronts another <u>one-eyed</u> ←✲→. Heisenberg intrinsic mono eyed reads <u>from to right</u> and sees 495 then, contacts the 4 then 9 and 5 and again sees 495. You read from left to right. Mono reads from center – both ways.

☼ 594◄—— ✲ ——►495☼

All the figures on both charts should be drawn upon one chart. (Two charts were for clarity.

Being that space is constant at the weesmall realm of the Heisenberg Zone, Dimensions and proportions are observed "relative to the observer". This phenomenon falls under Werner Heisenbergs Uncertainty Combined actions, at where the diagonal current crosses over through the 49.5 midpoint on the CAP chart, rotate the spin of the proton. Its structure is much like the less massive pion illustrated on page 14.

The portrayal of a one eyed Heisenberg may be "better understood" by realizing that space, being the vacant place that constitutes a position, possess a natural *apprehensive* characteristic endowed within. Momentum is "apprehended" by the position in space like as if position is female and the male is momentum. (If Sigmund Freud h ad known this position – momentum manner, of the way position apprehends the probe of momentum, he would have changed his idea that "everything is based on sex, like the female being apprehensive to the male approach", on down to the onstart of everything. Space itself is the origin of the one eyed Heisenbergs way of observation, Aristotles postulate, that everything begins to grow starting with one and proceeds to grow in a consecutive order as 1-2-3-4 etc., must be modified so that it begins, in growth, from zero and then proceeds consecutively. Therefore, another aspect of how observations are originated at the onstart position of the Heisenberg zone arises. This aspect rearranges dimensions and proportions into consecutive order as: 4356 is observed, from the initial position in space into consecutive order in these kind of arrangements.

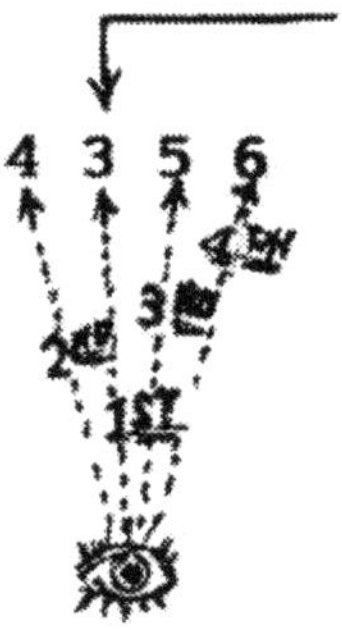

Initial observation begins with the least proportion.

4 3 5 6

1st 2nd 3rd 4th

Three, being the smallest, is apprehended first then four and five, arranged in a consecutively progressive order observed as 3456. A few other examples of the numerous ways that momentum and position initially form together are:

43+56=99 21+78=99 87+12=99 87+12=99

3456 34 4356 43 2178 21 8712 99 43+56=99

56 56 78 ↓ 21+78=99

99 99 99 99 99 99 99 (field maximum)

The exchange and transfer of momentum within an atom occurs in many ways. A hydrogen atom has only one proton and does not require a neutron that neutralizes the repelling forces that exist between two protons like are in hydrogen. The following charts display the 7X7 pion gaining 15 positions that make it into a 8X8 system that figures in the exchange process of the pion-neutron cooperation that transfers force.

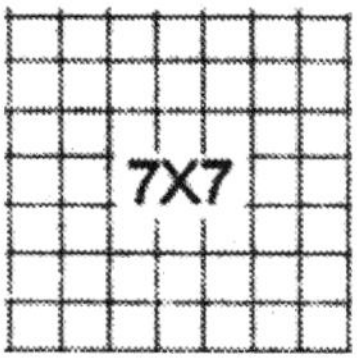

7X7

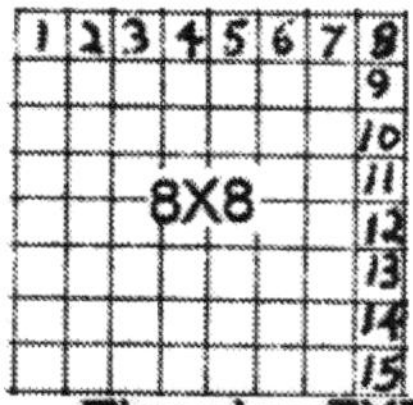

1	2	3	4	5	6	7	8
							9
							10
							11
							12
							13
							14
							15

8X8

1-3 moves

0 1 2 3 4 5

0	47	30	49	32	15	62	17
29	50	45	2	61	18	13	34
46	1	48	31	14	33	16	63
51	28	3	44	19	60	35	12
4	43	24	55	8	39	20	59
27	52	7	40	23	56	11	36
42	5	54	25	38	9	58	21
53	26	41	6	57	22	37	10

rows-columns-dia= 252

THE CHART TO RIGHT) → REFORMS FROM THE ABOVE AS →

THE ROWS & COLUMNS ON THE CHARTS BOTH TOTAL 252

51	60	3	12	19	28	35	44
13	2	61	50	45	34	29	18
52	59	4	11	20	27	36	43
10	5	58	53	42	37	26	21
54	57	6	9	22	25	38	41
8	7	56	55	40	39	24	23
49	62	1	14	17	30	33	46
15	0	63	48	47	32	31	16

The pion 7X7 chart [see 168 chart on page 20 gains 15 positions, 8 across top and 7 down side). The gain forms into a 8X8 isotropic <u>angular momentum</u> chart shown herein. Follow the angular momentum 1 over 3 down consecutively. The graph ☼ to the left of 8X8 chart is an example to show how to proceed to follow the 1-3 moves.

see page 31 (one over three down) for description of angular momentum shown in circles. Follow the angular momentum through entire chart and see that it touches all of the 4 positions. Then, the 8X8 chart reforms, preparing to shed the 15 positions it added onto the 7X7 pion. The reason that this happens is because the 8X8 has occupied more than 50% of the 10X10 field area of 100 positions.

Notice that the angular momentum (1 to 3) within the upper chart is different than the angular momentum in the lower chart. The first segment of the lower chart moves in a 1 to 3, starting with 51 to 61 instead of 1 to 3. The momenta in the lower chart has increased its difference by <u>10</u> (51 to 61). Also note that the second segment down has a difference of 6 (52 to 58). The upper system has spun its angular momenta in a way that makes it able to expel the extra 15 positions which shall be displayed next at top of column 8. What has happened is that the 8X8 (that glues onto the 7X7 pion) changes its momenta in order to be able to expel the extra 15 positions via the result of the change into the new formed 8X8 chart.

The chart, displayed below, starting with 51 then 60 across is the same system as the lower chart.

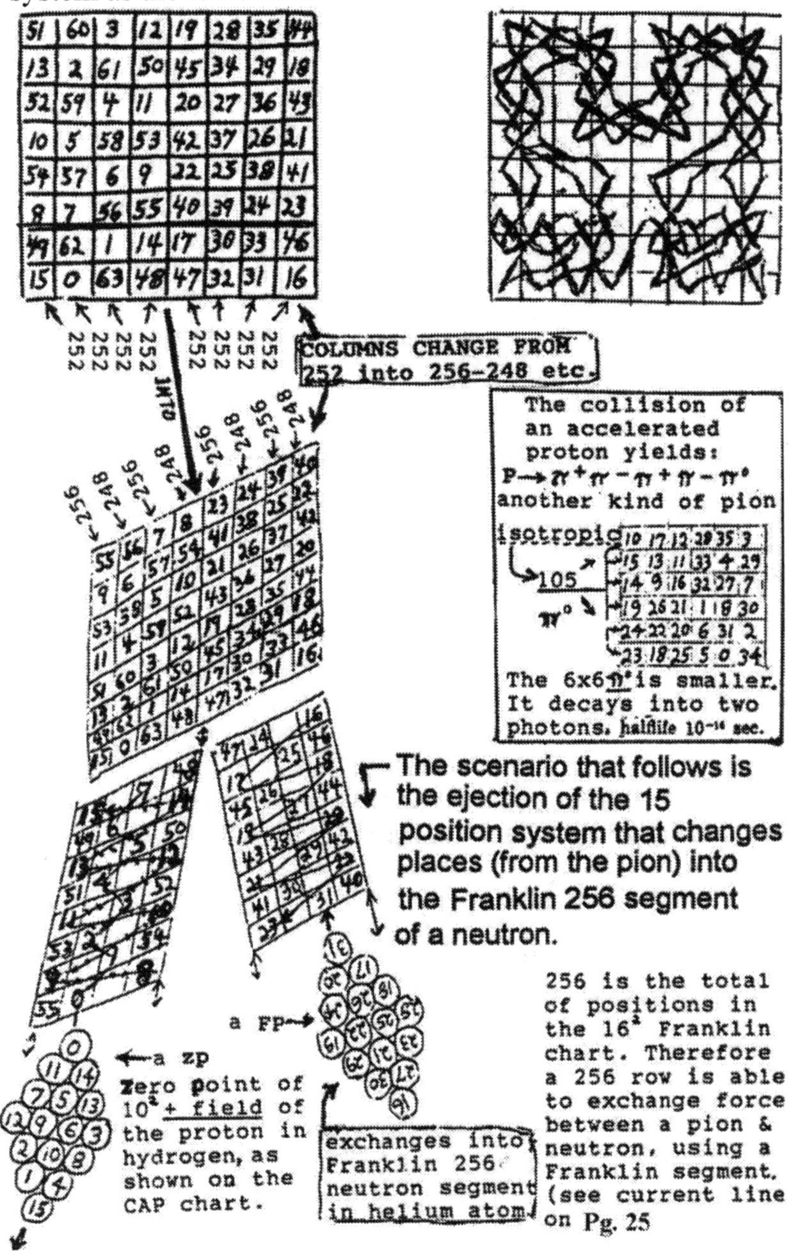

The explanation of ZP and FP that is in the Franklin chart displayed below.

AP = A Zero-point activated proton

FP = A proton that has the pion that exchanges a system that contains no zero in its make-up.

FRANLIN FORCE CHART

137-	1.	120 +256	Hydrogen	1ZP (no F)
	2.	376 +256	Helium	1ZP + 1FP
	3.	632 +256	Lithium	1ZP + 2 FP
	4.	888 +256	Beryllium	1ZP + 3FP
	5.	1144 +256	Boron	1ZP + 4FP
	6.	1400 +256	Carbon	1ZP + 5FP
	7.	1656 +256	Nitrogen	1ZP + 6FP
	8.	1912 +256	Oxygen	1ZP + 7FP
	9.	2168 +256	Fluoride	1ZP + 8FP
	10.	2424 +256	Neon	1ZP + 9FP

Every atom contains 1 zero point proton exchange (ZP). 255 is the total momenta in a 256 system [256-119=137]. 119 (not 120) can't add 1 zeros.

On pg 40 The zero point (done in circles) is ejected from the 8X8 field system and enters the Franklin field system. The way that it maneuvers into the Franklin field (a segment of the neutron) is illustrated by the current lines drawn on the Franklin chart displayed Page 25.

The zero point system contains 16 positions. It contains an empty position. Only 15 positions contain momentum. The isotropical makeup gives each row a total momenta of 30. Four rows of 30 make the total momenta 120 [4 times 30=120].

The zero point system is #1 on the Franklin Force Chart that is displayed on pg. 41 on the right side of the exchange scheme.

Every atom requires the zero point (0 thru 15) system as the initial additional proton added to the nucleus as the atoms increase in size up the periodic table of elements. The reason that this zero point (120 momenta) is necessary to add another proton to form a larger atom, is that the zero is needed to onstart the larger atom. Note that the ejected twin, from the above pion exchange, begins with 16 and has no zero in its system. See helium on the above Franklin Force Chart. Helium contains one zero-point system and the next larger 4X4 (twin) system in its nucleus, plus 1 Franklin 256 positional chart (of the segment of a neutron) that figures in the exchange process. The charting of momentum in an activated zero-point system is displayed below. It is the *opposite* directed portrayal of the reciprocating system that is displayed at

A zero-point system manipulates its momentum through 64 positions housed in *each* of the leftward and rightward reciprocated motion. 16X16=256.

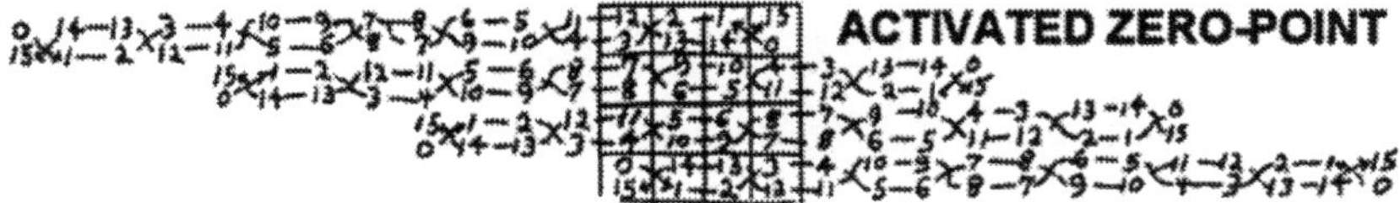

The chart, displayed next, shows the manner that pions exchange force into a Franklin segment of a neutron as each atom, that is listed on the Periodic Table of Elements, increase in size. Atoms employ different increments of exchange systems. Their capacities are charted next.

The small charts (done in gridlocked squares), that are displayed beside the *Pion Exchange Chart*, mathematically illustrate the momentum changing position from the 8X8 pion scheme as is displayed on page 40.

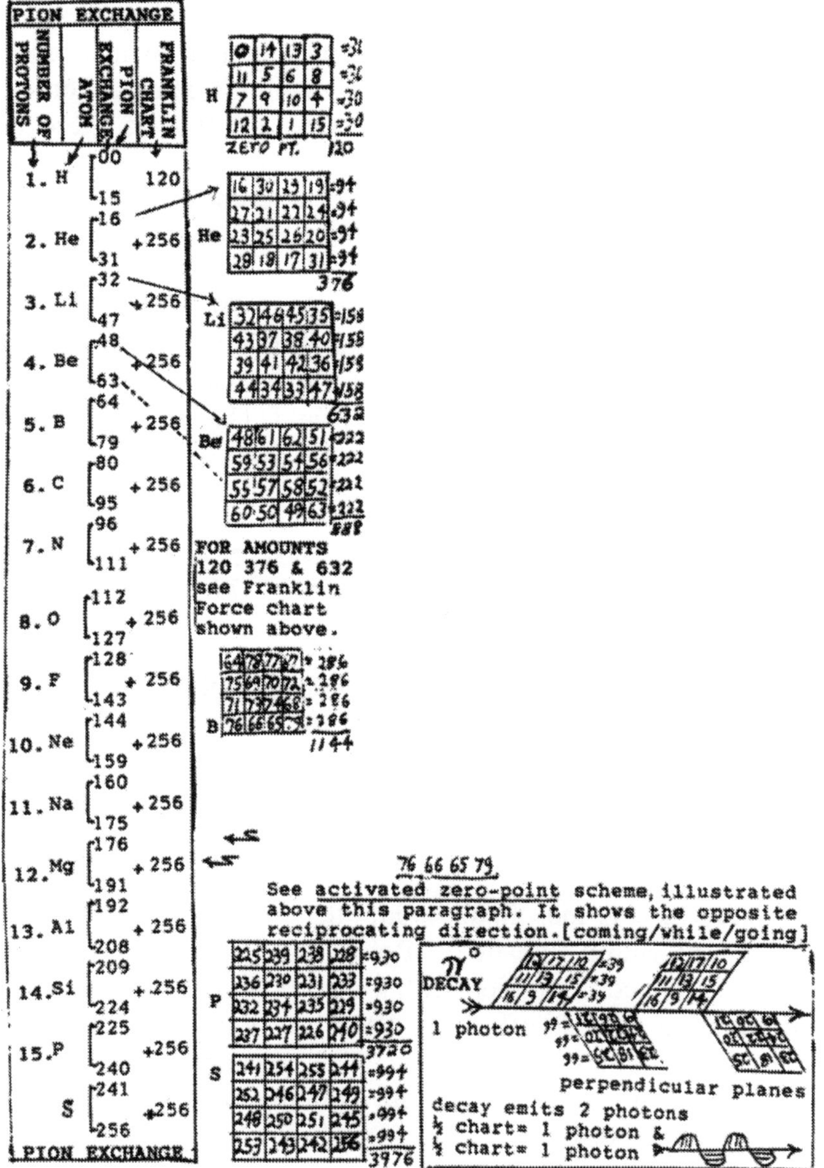

*See Franklin Chart at the top of page 25 that shows the current entering the neutron segment. The current lines represent the zero-point (0 – 15) exchange of a hydrogen pion. Each pion, that exchanges force, attains a higher capacity as the consecutively arranged protons are situated in the nucleus. Thus, the frequency of the pion becomes wider as can be depicted by starting at 88 on the top of Franklin chart on page 25, and proceeding down to 89 and back across to 90 and on back to 91. The consecutively charted systems, that exchange in a Lithium atom are a zero-point (ZP) which begins at 0 and proceeds to 15. The next exchange action is charted from 16 to 32 and fits into a helium atom along with the 0-to-15 (zero point system). The final proton, of the three protons that make up the nucleus of Lithium, spins off the pion that reaches the capacity that fills the field, this is charted from 32 to 47.

Boron, being exchanged by the chart 64-to-79, is the atom that is able to contain its positive (+) configuation within the bounds of the 100 by 100 (10,000 positional) CAP Chart. The configuration of carbon extends *off* the chart into a larger field. The larger field figures in the DNA molecule structure, as is revealed in the book *Price Principles Of The Heisenberg Zone*.

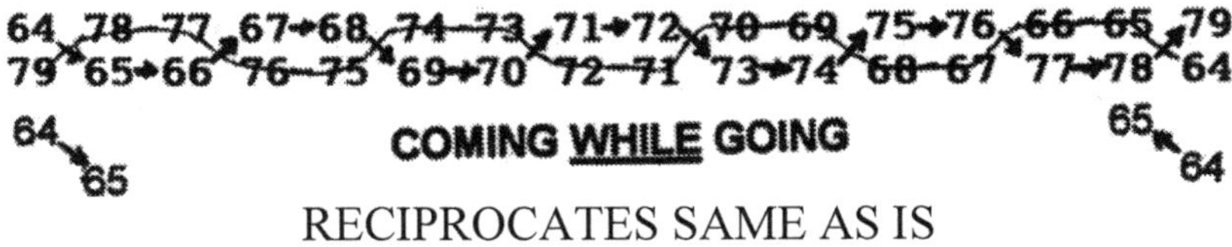

RECIPROCATES SAME AS IS
DISPLAYED ON PAGE 42

See activated zero-point scheme, illustrated above this paragraph. It shows the opposite reciprocating direction. [coming/while/going]

Phosphorus is the final atom that contains its positive+ configuration within its field. Boron was the final atom that maintained its + configuration within the 100X100 field. The carbon + configuration extends off of the 100X100 field, and joins into the forming of the amino acids in DNA. This phenomenon is illustrated on the large DNA chart that is displayed in the *Price Principles of the Heisenberg Zone* book.

Sulfur is the final atom that exhanges strong force between a pion and the Franklin segment of a neutron. The reason is because the sulfur pion exchange capacity is 256 (chart limit).

Douglas M. Price

SECTION II

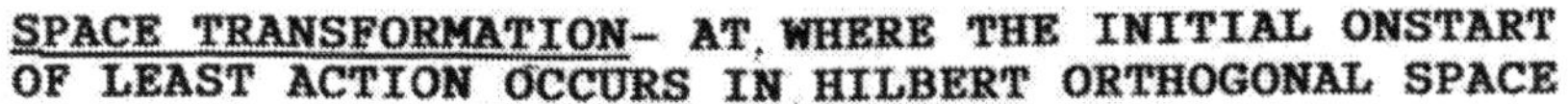

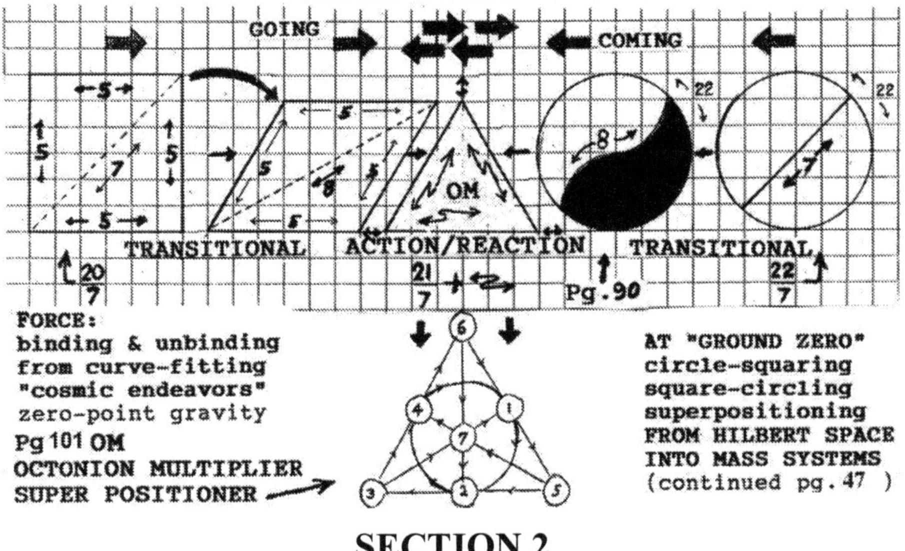

SECTION 2

THE FOLLOWING DATA IS A PREVIEW OF
THE WRITINGS THAT WILL BE EXPLAINED
LATER.
*See

pg. 2-THE ENCLOSED CAP CHART ILLUSTRATES THE POSITION AND MOMENTA FUNCTIONING AT THE POSITIVE+ CHARGED AREA IN THE NUCLEUS OF THE TEN ATOMS-Hydrogen-Helium-Lithium-Beryllium-Boron-Carbon-Nitrogen-Oxygen-Fluorine and Neon. (Order of Periodic Table).

The structure of the chart is a two-dimensional graphing scheme of the positive+ charged field surround of the nucleus. The Cap Chart, as displayed herein, only depicts the Hilbert orthogonal space background that has the *potential* to house mass when activated. The activation process, that causes the Hilbert potential area to structure into mass, happens when the horizontal field area is approached by momenta that comes from the opposite direction then what is displayed on the CAP CHART. The result is like taking a bit of some time area of the past and adding some of the time area that comes from the future and combining it into a system that houses a "*presence*" of time. The combining endeavor brings mass into existence, being in the real world that we observe.

---Its about *time* that mankind learns exactly what *time* is ---

The following paragraph will define the meaning of space and time in a manner that will not require a rockologist to be able to comprehend space-time.

A geometric drawing of space is not possible, because space is non-existent. Space has no presence until something *happens* to a place in space. If *nothing* happens, then space and time remain constant.

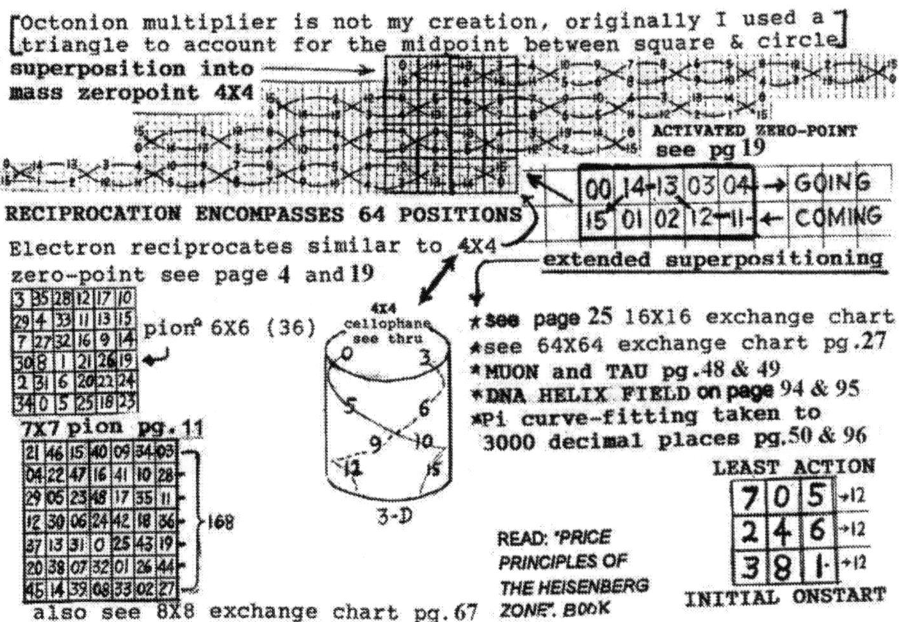

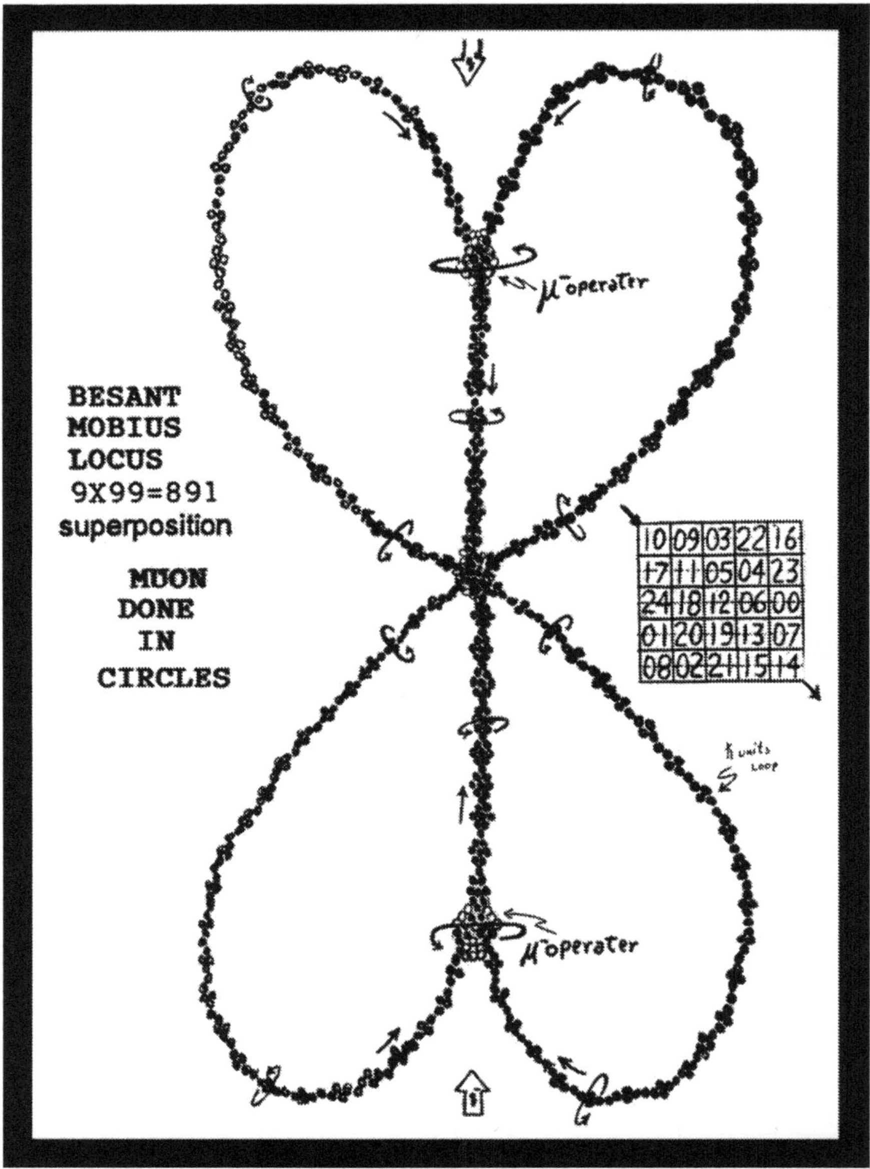
BESANT
MOBIUS
LOCUS
9X99=891
superposition
MUON
DONE
IN
CIRCLES
μ^-operater
μ^-operater
10 09 03 22 16
17 11 05 04 23
24 18 12 06 00
01 20 19 13 07
08 02 21 15 14
units
loop

Electron Field Accelerated 99 × 99 = 9801

TAU PARTICLE

SUPERPOSITIONED **ZERO POINT**

0000	0099	0198	0297	0396	0495	0594	0693	0792	0891
0990	1089	1188	1287	1386	1485	1584	1683	1782	1881
1980	2079	2178	2277	2376	2475	2574	2673	2772	2871
2970	3069	3168	3267	3366	3465	3564	3663	3762	3861
3960	4059	4158	4257	4356	4455	4554	4653	4752	4851
4950	5049	5148	5247	5346	5445	5544	5643	5742	5841
5940	6039	6138	6237	6336	6435	6534	6633	6732	6831
6930	7029	7128	7227	7326	7425	7524	7623	7722	8721
7920	8019	8118	8217	8316	8415	8514	8613	8712	8811
8910	9018	9108	9207	9306	9405	9504	9603	9702	9801

ELECTRON
9X11=99 ← **zeropoint** → **MUON**
99X99=9801

99	98	97	96	95	94	93	92	91	90
89	88	87	86	85	84	83	82	81	80
79	78	77	76	75	74	73	72	71	70
69	68	67	66	65	64	63	62	61	60
59	58	57	56	55	54	53	52	51	50
49	48	47	46	45	44	43	42	41	40
39	38	37	36	35	34	33	32	31	30
29	28	27	26	25	24	23	22	21	20
19	18	17	16	15	14	13	12	11	10
09	08	07	06	05	04	03	02	01	00

000	009	018	027	036	045	054	063	072	081
090	099	108	117	126	135	144	153	162	171
180	189	198	207	216	225	234	243	252	261
270	279	288	297	306	315	324	333	342	351
360	369	378	387	396	405	414	423	432	441
450	459	468	477	486	495	504	513	522	531
540	549	558	567	576	585	594	603	612	621
630	639	648	657	666	675	684	693	702	711
720	729	738	747	756	765	774	783	792	801
810	819	828	837	846	855	864	873	882	891
900	909	918	927	936	945	954	963	972	981

Pi (π) - TAKEN TO 3000 DECIMAL PLACES

1415926535 8979323846 2643383279 5028841971 6939937510 5820974944 5923078164 0628620899 8628034825 3421170679
8214808651 3282306647 0938446095 5058223172 5359408128 4811174502 8410270193 8521105559 6446229489 5493038196
4428810975 6659334461 2847564823 3786783165 2712019091 4564856692 3460348610 4543266482 1339360726 0249141273
7245870066 0631558817 4881520920 9628292540 9171536436 7892590360 0113305305 4882046652 1384146951 9415116094
3305727036 5759591953 0921861173 8193261179 3105118548 0744623799 6274956735 1885752724 8912279381 8301194912
9833673362 4406566430 8602139494 6395224737 1907021798 6094370277 0539217176 2931767523 8467481846 7669405132
0005681271 4526356082 7785771342 7577896091 7363717872 1468440901 2249534301 4654958537 1050792279 6892589235
4201995611 2129021960 8640344181 5981362977 4771309960 5187072113 4999999837 2978049951 0597317328 1609631859
5024459455 3469083026 4252230825 3344685035 2619311881 7101000313 7838752886 5875332083 8142061717 7669147303
5982534904 2875546873 1159562863 8823537875 9375195778 1857780532 1712268066 1300192787 6611195909 2164201989

3809525720 1065485863 2788659361 5338182796 8230301952 0353018529 6899577362 2599413891 2497217752 8347913151
5574857242 4541506959 5082953311 6861727855 8890750983 8175463746 4939319255 0604009277 0167113900 9848824012
8583616035 6370766010 4710181942 9555961989 4676783744 9448255379 7747268471 0404753464 6208046684 2590694912
9331367702 8989152104 7521620569 6602405803 8150193511 2533824300 3558764024 7496473263 9141992726 0426992279
6782354781 6360093417 2164121992 4586315030 2861829745 5570674983 8505494588 5869269956 9092721079 7509302955
3211653449 8720275596 0236480665 4991198818 3479775356 6369807426 5425278625 5181841757 4672890977 7727938000
8164706001 6145249192 1732172147 7235014144 1973568548 1613611573 5255213347 5741849468 4385233239 0739414333
4547762416 8625189835 6948556209 9219222184 2725502542 5688767179 0494601653 4668049886 2723279178 6085784383
8279679766 8145410095 3883786360 9506800642 2512520511 7392984896 0841284886 2694560424 1965285022 2106611863
0674427862 2039194945 0471237137 8696095636 4371917287 4677646575 7396241389 0865832645 9958133904 7802759009

9465764078 9512694683 9835259570 9825822620 5224894077 2671947826 8482601476 9909026401 3639443745 5305068203
4962524517 4939965143 1429809190 6592509372 2169646151 5709858387 4105978859 5977297549 8930161753 9284681382
6868386894 2774155991 8559252459 5395943104 9972524680 8459872736 4469584865 3836736222 6260991246 0805124388
4390451244 1365497627 8079771569 1435997700 1296160894 4169486855 5848406353 4220722258 2848864815 8456028506
0168427394 5226746767 8895252138 5225499546 6672782398 6456596116 3548862305 7745649803 5593634568 1743241125
1507606947 9451096596 0940252288 7971089314 5669136867 2287489405 6010150330 8617928680 9208747609 1782493858
9009714909 6759852613 6554978189 3129784821 6829989487 2265880485 7564014270 4775551323 7964145152 3746234364
5428584447 9526586782 1051141354 7357395231 1342716610 2135969536 2314429524 8493718711 0145765403 5902799344
0374200731 0578539062 1983874478 0847848968 3321445713 8687519435 0643021845 3191048481 0053706146 8067491927
8191197939 9520614196 6342875444 0643745123 7181921799 9839101591 9561814675 1426912397 4894090718 6494231961

FURTHER ACCELERATED FIELDS ARE HELIX FORMS

RNA---99X999 = 98901
DNA INITIATOR(starter)----99X9999 = 989901 see page 95.
see large DNA chart procured at Pennick Scientific Enterprises P.O. BOX 382 Carnegie Pa 15106.
The final field discussed in this book is the immense pi decimal expansion field. An endless field taken to only 3000 decimal places. It transposes ALL the characteristics across the amino acid sets in the 2 stranded helix

GRAVITY

ACTION OCCURRING AS IF VIEWED AT NANO-SECOND

The surrounding area of atoms that houses negative and positive force activity (electron cloud). The predominately negative force lessens as it nears the positive nucleus of the atom at the quantum area

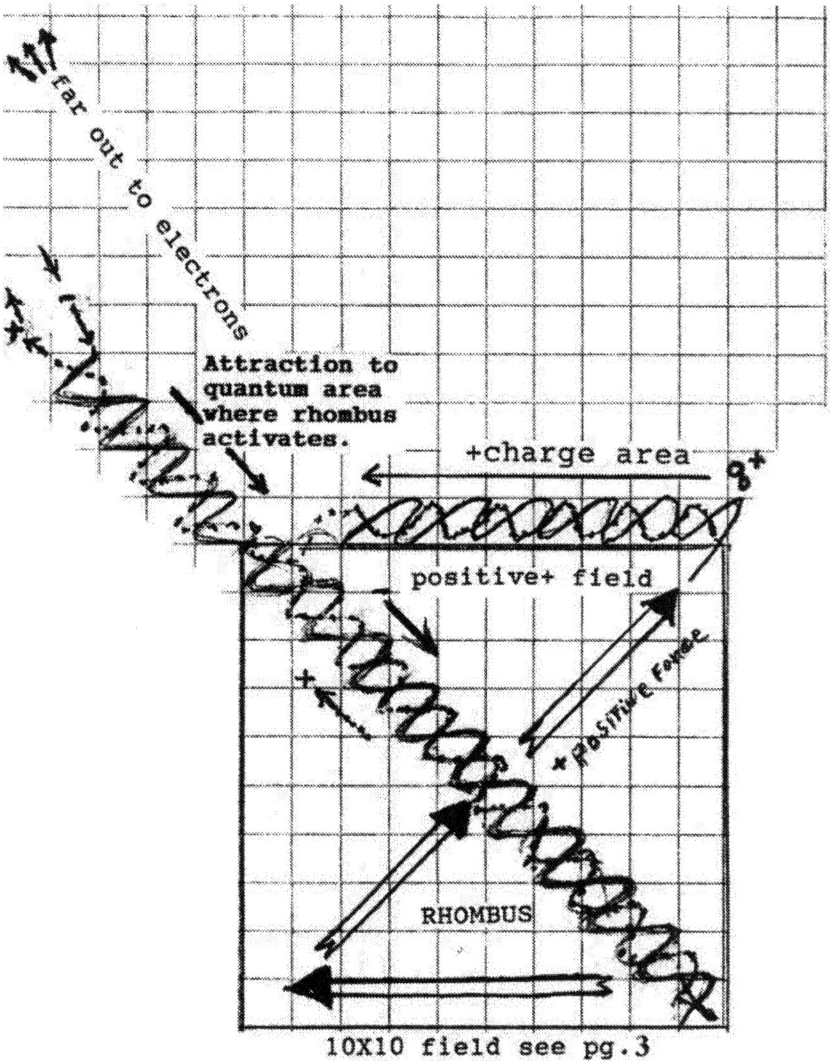

far out to electrons
Attraction to quantum area where rhombus activates.
+charge area
positive+ field
+positive force
RHOMBUS
10X10 field see pg.3

-PUTTING IT ALL TOGETHER-

PROOF THAT THE 4X4 CHART IS THE SCHEME THAT INITIATES THE "SPACE INTO MASS" SUPER POSITIONING. THE ZERO-POINT ONSTART SYSTEM OF MASS FORMS.

DIAGONAL 7
20 PERIMETER
22/7=3.142857142857142
20/7=2.857142857142-----
infinite repeater
DIAMETER
22/7
22 CIRCUMFERENCE
00 14 13 03 =30
11 05 06 08 =30
07 09 10 04 =30
12 02 01 15 =30
see pg.98

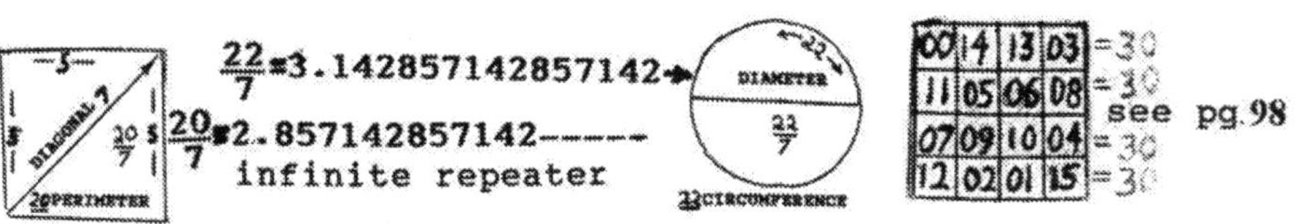

In order for energy to develop momentum into position that creates a system able to *exist*, it must be relative to the viewer from all directions. See 4X4 zero-point chart. (above) If the perimeter comes into being from one side to the other it violates the Law of Conservation of Momentum because it requires time to move from one position to another. Time changes things. (Heisenberg Uncertainty Principle). HOW IT HAPPENS:

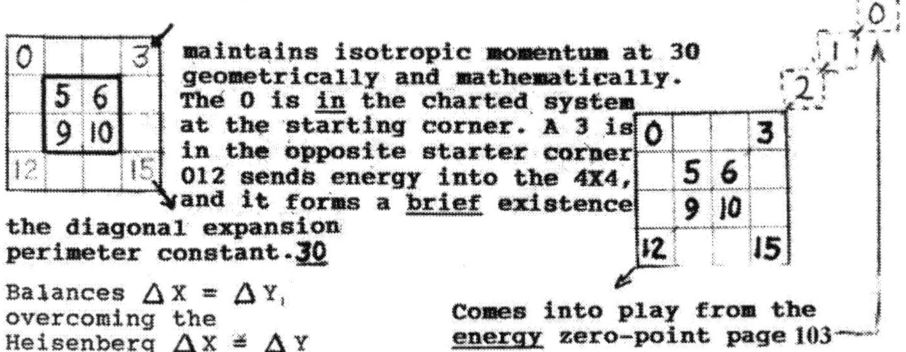

Maintains Conservation of Momentum law

The 4X4 charted system focuses (balances) the Heisenberg Uncertainly Principle. The pion 7X7 charted system *decay mode* separates into 3X3 and 4X4 charted systems. see pg. 21

SUMMARY SECTION I

Most of the charts in section I illustrate the momentum in the charted systems as if stopped at the nano-second where the system is isotropically balanced. A isotropically arranged system changing momentum & position is charted so that the row, column and diagonal amounts are the same, relative to observation from all directions. The result gives the charted systenm a brief lifetime. The midpoint of the chart system is where it overcomes the Heisenberg ΔX=ΔY Uncertainty problem, by balancing the system into focus as:

→07 2 5 4 3 6 1 8 → {at 4,
←8 1 6 3 4 5 2 7 0← Δ x = Δ y

If the sub-atomic system could be viewed in motion (momentum changing position) it would appear somewhat like a bunny rabbit "kicking inside a cloth sack. The sack would appear as if it was jutting in & out in various directions. (see a chart portrayed as stopped nano-second showing momentum jutting from the system on page 47.

Since there is no name for t his jutting scenario, I will take the liberty of naming it a bunner and will also name the system with the jutting bunner a "gunning system". Bunnering is the phrase that explains the overall action and re-action view of the system as time lapses while momentum is changing position. The 4X4 charts on page 54 are all isotropically arranged (momentum/position row & column total) Each chart illustrates a single phase of the zero-point base of the "bunning" system as if stopped at one nanosecond jut. When the zero-point bases of the 16 bases as on page 54 are super positioned together into a more massive system form, the result is the structure of the 16 X 16 Neutron-Franklin chart on page 25. Note the current changing position via 4X4 increments. When the 16X16 Franklin is energized into super position it becomes a neutron as displayed on page 27. Then super-position to a muon as the besanting system on page 48 and further super positioning can lead up to the helix DNA. The charts on page 54 only show ½ of the action/reaction of the zero-point system. In order to capture the other half, fill in the difference into each position of the chart, so that every position contains 15. Example:

15
00

14	13
01	02

etc.

This effort reveals 15 additional isotropic charts.

5	2	9	14				0	13	14	3				11	5	6	8
8	15	4	3				11	6	5	8				12	2	1	15
6	1	10	13				7	10	9	4				0	14	13	3
11	12	7	0				12	1	2	15				7	9	10	4
7	10	13	0				5	2	9	14				3	13	14	0
12	1	6	11				8	15	4	3				8	6	5	11
2	5	8	5				6	1	19	13				4	10	9	7
9	4	3	14				11	12	7	0				15	1	2	12
0	13	6	11				9	14	2	5				3	14	13	0
14	3	8	5				15	4	8	3				8	5	6	11
9	4	15	2				0	11	7	12				4	9	10	7
7	10	1	12				6	1	13	10				15	2	1	12
14	9	2	5				0	12	11	7				12	1	2	15
3	4	15	8				3	15	8	4				7	10	9	4
13	10	1	6				13	1	6	10				11	6	5	8
0	7	12	11				14	2	5	9				0	13	14	3
0	14	13	3				0	4	15	11				11	7	0	12
11	5	6	8				9	13	2	6				8	4	3	5
7	9	10	4				14	10	5	1				6	10	13	1
12	2	1	15				7	3	8	12				5	9	14	2

The next one inch of space on this paper is left blank, in order to illustrate the non-geometry and non-math scheme of undisturbed (non-existent) space. ***ZERO SPACE IS NON-MEASURABLE.***

[NOTHINGLESS]

It requires motion for something to happen in space (motion – takes time) like 0-to-01. Zero is even, one is odd. Time is a non-existent and non-clockable constant until **something disturbs constant space**. A simple vector motion takes time to happen. 00 then time can be clocked as such. Time = distance/rate. ↘
01

The purpose of this introduction is to reveal the math and geometry of the structural details of a proton and its positive field surround. Therefore, the sub-quark forms etc. that are displayed in the next named, *The Price Principles of the Heisenberg Zone* are omitted. A description of Plancks energy unit follows later.

A stable system, like a proton, sets up in psace different then a proton (shown below) because it is housed in its place in space which sets up as: **A STABLE *PRESENCE* OF TIME**.

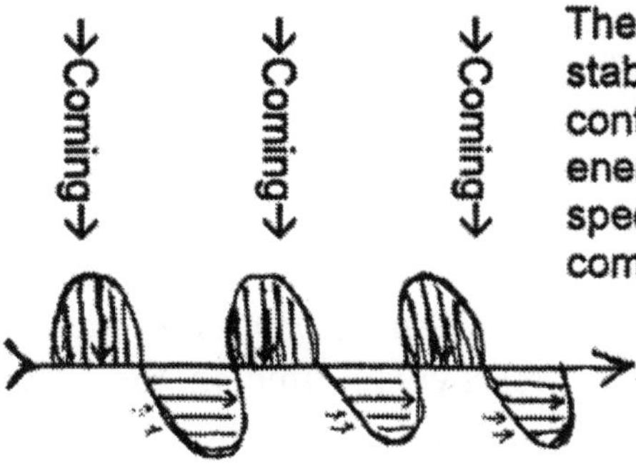

↓Coming↓ ↓Coming↓ ↓Coming↓

Going → Going → Going
PHOTON SCHEME

The photon is not housed in a stable place in space. It contains momentum and energy. It maneuvers at light speed toward the future. The coming/going frequencies are set by electron acts. *SOMETHING FOR YOU TO PONDER* Proton emits electron from sun electron goes X years while comes. Proton stops emitting electron the electron beam comes (returns) while going

for X years. Now what do you see looking at starlight? *SEE IT AND SAY IT LIKE IT IS!* How time is established in a system, in space. We can make a clock with a pendulum, swinging back and forth ----$h/2\pi$. A swing forth is one π *going*, and a swing *coming* back is 1π ($h/2\pi$). We can look at the moon phase month to month, that becomes from dark to brighter. There are many ways to measure time. The essence of time, at the onstart from zero-space that is established in a fundamental system, requires a space transformation, which makes it capable of having an *existence*.

The following diagrams show, in steps, how the positive field of a proton is established as a stable particle in zero-space.

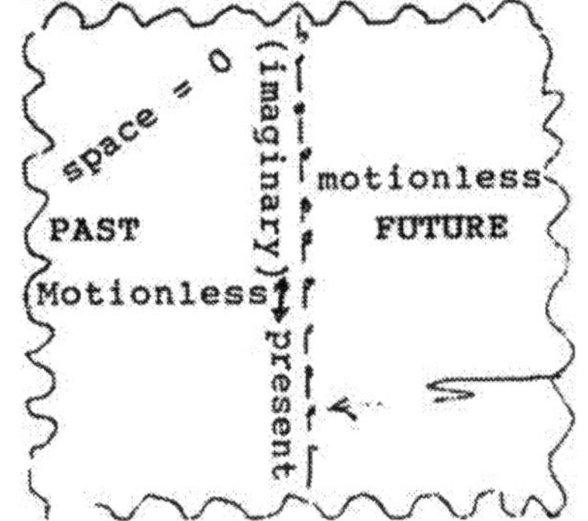

Imaginary boundary
·Boundless
·Non-existent
·Its not there and non-measurable
·Has no presence
·No geometry
·No Math scheme
·The dotted line (also imaginary) represents the **A potential place in Hilbert** PRESENT which is a ***zero-space-time constant***. constant.

The display (pg. 55) is nothingless which possesses a potential to be something when disturbed (probed etc). This display is to illustrate what space is like before the onstart that structures mass forms. The proton field, (displayed soon) results from this onstart potential.

In order for *something* to exist, it must have a "presence". It requires a *place* in space that "houses" the *something*. Motion is necessary in order to establish a presence of time in the place in space. Instead of being motionless, like the above shown scheme, a presence of time replaces the potential time constant (imaginary dotted line), to include some of the past with some of the future creating a thing that exists. The next illustration portrays the potential place in space *activated* into existence (Time established).

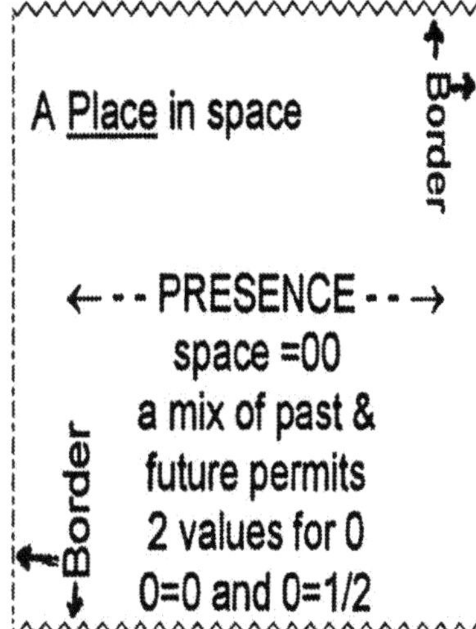

AN ELECTRON REQUIRES THE "PRESENCE" OF MASS IN ORDER TO EXIST. *ANYTHING* THAT EXISTS REQUIRES THE "PRESENCE" OF TIME AND CONSEQUENTLY, TIME ALSO REQUIRES A "PLACE" IN SPACE.
(width). Time, at the onstart construction, *always* begins at zero, because zero is mathematically positioned as the initial forecost place in space. So much for the geometry. A math scheme sets up in the geometrical area that has established a place in space. The scheme (shown later) mathematically graphs co-junction of past and future in a given area.

Instead of the single imaginary dotted line in the above drawing that depicts time as constant, two dotted lines *spaced apart* from each other, form a place in space that involves co-junction of some past with some future. The dotted lines are distanced apart from each other. It takes time to travel from line to line. WHAT IS TIME?

TIME is a co-partner of motion. To travel between two points requires motion. Consider time --- to be the *rate* of the motion (speed) that transfers between two locations.

Our minds base the rate upon the motion of clocks, which are programmed by the rotation period of earth --- 24 hours minutes, seconds etc. In order to prove t his definition of time, simply remove the sun and planets and all mass from space, and see then, that there *is no* time in zero space. Time is constant as much. Modify the theory of the universe.

The next paragraph deals with the aspects of photons, then, knowing the structure and functioning of photon, it will be possible to illustrate how a stable system is establish in zer-space.

A light photon travels in a straight line at about the rate of 186,274 miles per second. If a photon could travel in a curved path, it would take about ¾ of a second to circumnavigate earth, released from your flashlite. ¾ of a second is about the time it takes to snap your finger. Photons contain energy --- motion --- direction. See photon scheme picture on page 55.

A photon is derived from the excess energy that is ejected from the electron field. An electron's field consists of going and coming momentum changing in each of its 100 positions. A diagram of the locus of an electron is on page 9. Note the twisting field that ejects the excess energy in perpendicular planes.

The hydrogen atoms positive+ field area is displayed at the lower lefthand corner on the CAP chart. It is housed in a square form consisting of 100 positions (10X10). In order for it to be activated into existence, a coming/while/going math scheme must be applied, mathematically structured as:

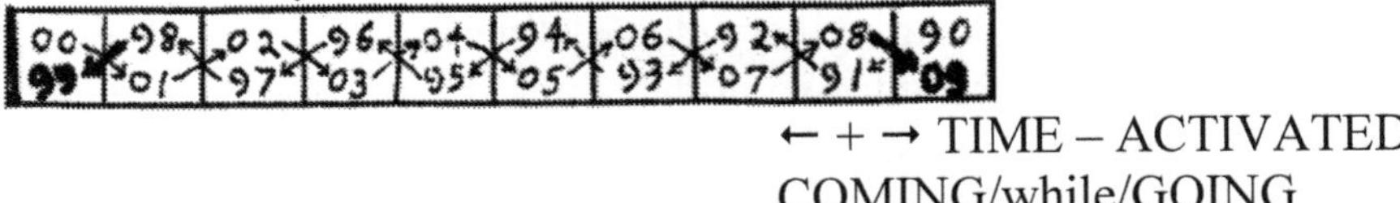

← + → TIME – ACTIVATED
COMING/while/GOING

↑
THIS IS THE TOP ROW OF THE 100 POSITIONAL FIELD SHOWN BELOW:
Compare this scheme to the bottom graph on page 4.

Activated scheme of the top row of the hydrogen atom *positive+* field (CAP CHART).

When activated, the top row of the potential Hilbert space field has an oppositely directed math scheme manipulating horizontally. The entire 10X10 system creates a "presence" which brings it into existence,

99 / 00	98 / 01	97 / 02	96 / 03	95 / 04	94 / 05	93 / 06	92 / 07	91 / 08	90 / 09
89 / 10	88 / 11	87 / 12	86 / 13	85 / 14	84 / 15	83 / 16	82 / 17	81 / 18	80 / 19
79 / 20	78 / 21	77 / 22	76 / 23	75 / 24	74 / 25	73 / 26	72 / 27	71 / 28	70 / 29
69 / 30	68 / 31	67 / 32	66 / 33	65 / 34	64 / 35	63 / 36	62 / 37	61 / 38	60 / 39
59 / 40	58 / 41	57 / 42	56 / 43	55 / 44	54 / 45	53 / 46	52 / 47	51 / 48	50 / 49
49 / 50	48 / 51	47 / 52	46 / 53	45 / 54	44 / 55	43 / 56	42 / 57	41 / 58	40 / 59
39 / 60	38 / 61	37 / 62	36 / 63	35 / 64	34 / 65	33 / 66	32 / 67	31 / 68	30 / 69
29 / 70	28 / 71	27 / 72	26 / 73	25 / 74	24 / 75	23 / 76	22 / 77	21 / 78	20 / 79
19 / 80	18 / 81	17 / 82	16 / 83	15 / 84	14 / 85	13 / 86	12 / 87	11 / 88	10 / 89
09 / 90	08 / 91	07 / 92	06 / 93	05 / 94	04 / 95	03 / 96	02 / 97	01 / 98	00 / 99

as such. The bottom row of the field appears, when it is activated, as:↘

90	08	92	06	94	04	96	02	98	00
09	91	07	93	05	95	03	97	01	99

In honor and respect for the research work done by String Theorists, I'll refer to the above coming/while/going charts as a bit like a binded string (the word string isn't correct).

HOW THINGS COME INTO EXISTENCE ----

What comes first, the chicken or the egg? (Will answer soon).

Mankind has attempted to decipher the mysteries of the cosmos that account for the presence of sub-atomic particles and the unknown mathematical and geometrical details of their structure. I am presenting this CAP chart and data, that will reveal most of the long sought questions that are concerned with how things occur at the onstart of mass from space.

This article is written in a manner so that it doesn't require a Rocket sci rockheadologist to understand it. ARISTOTLE proposed that (mathematically) everything starts at one and continues in growth in a consecutive order as 1 – 2 – 3 – 4 etc. O.K. maybe so, if *zero is placed before 1* to account for space. Therefore, "small precedes large". Eggs are small, chickens are large. So be it.

CONCERNING DIMENSIONS AND PROPORITONS – MOMENTUM AND POSITIONS

1941 at Princeton Unversity, three professors, Einstein, Bergman and Bargmann, were trying to work out a theory named the 4th dimension, so that it could fit into a 5th dimension theory proposed by Theodore Kazula. They believed that an electron was a point particle and attempted to find a scheme that would fit as a zero-point that formed the onstart of an electron particle. JOHN DIRAC (on was it Pauli?)

0	1
1	0

conjectured that an electron sort of appears in and out of space. He employed the use of this simple matrix form as the math.

The note about the two values that apply to zero is written on the *presence chart* displayed previsouly. Zero (0) has nothing to measure, so *0 = 0*. But, not so in the *presence* geometric scheme which combines past with future in a given place in space. The nothingless zero-space chart is displayed a bit above the presence chart and contains a nothingless value. Due to the combining of past with future making a geometric presence scheme, the value becomes 00 = ½. Nothingless zero space is 0 = 0 and when a presence occurs at the onstart of the structure of elementary mass forms, then 00 = ½. Notice that all of the charts in this article begin with 00.

There is another underlying math scheme that occurs during the very onstart of sub-atomic structures that must be presented. It concerns what happens in the area from zero space to one (0 --- 1). Things do not just pop into real amount numbers from zero-space. At the initial beginning observation action-reaction occurs which has a midpoint situation ½ way between real amount digits. Some examples are to follow a bit.

Space is instantaneous. [Put your mind to this]. A POSITION ANYWHERE IN EVERLASTING SPACE IS INSTANTANEOUS, because it is *already there*! Whatever occurs in a area in zero-space like direction --- momentum --- motion etc, requires time. Time comes and goes. The Hamiltonian Principle of Least Action happens at the realm of an *occupied* set of positions at the fundamental onstart of energy. This set of positions has a midpoint where the momentum comes from zero-space until it reaches the midpoint, then the momentum *goes* from the ½ way midpoint until it reaches the end of the system. Sir Issac Newton's "action re-action postulate applies to this (coming while going) scenario of space transformation that initiates action into a system in an area of instantaneous space. All space has the *potential* to structure mass.

|
→GOING TO ---*---COMING FROM ← {THIS SIGNIFIES THAT
MID THE POSITIONS IN THE
←COMING FROM-- * -- GOING TO → SYSTEM BALANCES
| THE MOMENTA AT
MIDPOINT.

The above shown scheme can be made into a line graph (displayed soon).

Fields, such as proton or electron surround area, are derived from the superpositioning of zero-point states. The field of atoms, that are arranged starting from hydrogen on up the periodic table of elements, increase in size due to the capacity of the accumulating 20/7 square-circuling that occurs (separated at 5.5 positions) up the diagonal of the CAP chart. Notice on the CAP chart, that one proton exist centered in the initial 10X10 field at 5.5 up along the diagonal. The single proton functions a single positive+ charge across the top of the 10X10 field. Then, proceeding up the diagonal of the Cap chart, another proton becomes centered at position 11 (2X5.5) in a 20X20 positional field which has the capacity to support 2 protons, one in the 10X10 hydrogen field and one in the 20X20 helium field. A single proton has the capacity of positive+ (going) current force, to set up a single electrons negative (coming) current force at far away from the protons positive field. Postive+ force is outward ↑
← →
↓

↓
going →, whereas, negative- force is inward *coming* → ← ←
↑

Following, are a few of the many 3-dimensional structural developments that occur, at the onstart of mass forms, that form a system from oo-space. These are the base for super positioning into larger systems.

A ling-graph scheme of a 2-dimensional virtual momentum form.

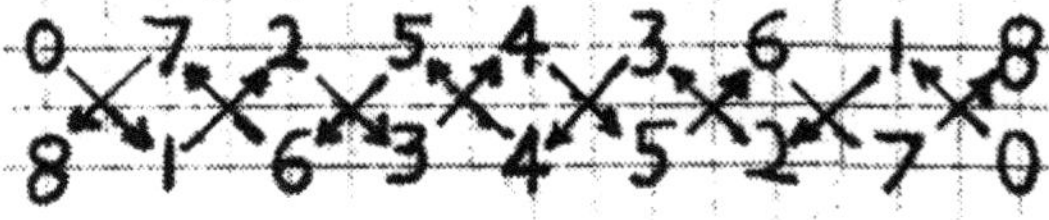

STRUNG TOGETHER RECIPROCATING COMING/GOING LINE GRAPH.

Maintains added total of 8 throughout the exchange

0 7 2
8 1 6 etc.
8 8 8

The important thing to know about a line-graph system is the timing sequence. The above shown line-graph (o thru 8) shows starting at zero and ***coming*** toward the midpoint at 4.

Then, going away from the midpoint to 8. The re-action sequence is opposite, -- → ***coming*** from 8 toward 4 (midpoint) then ***going*** away from 4 to zero. This equalizing action overcomes the *Heisenberg Uncertainty Principle* by focusing at the midpoint.

This line-graph style of *coming/while/going* current is how the current is tied together going from the helix strand A and coming back from strand B in the DNA molecule. A huge DNA chart depicting this activity is displayed in the book named "*The Price Principles of the Heisenberg Zone*". Later in this section a chart of a proton particle having 49.5 as a midpoint will be illustrated.

PYTHAGORAS (540 B.C.) was a Greek mathematician that worked and proved the fact that the square on the long side of a 3-4-5 right triangle equals the sum of the squares on the smaller sides, shown as:

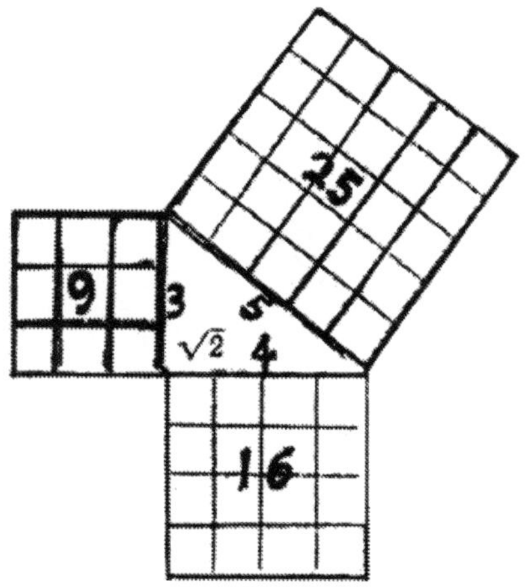

This 3-4-5 scheme, of a right triangle, is nature's "basic" system that is involved with the superpositioning of mass forms into larger mass forms. Hilbert orthogonal space is all evened. Therefore, I must point out that the planes on the 3 side and 4 side are not evened with the plane of the 5 side. It is nearly diagonal. The differentiation of planes causes the "spacing" of the Pythagoras system to become activated, by motion, into existence.

The 3^2 (9 positional) system is the zero-point system from which *energy* forms are superpositioned. The 4^2 (16 positions) systems is the zero-point system that superpositions into mas forms.

By adding the mathematics into Pythagoras right triangle squared system, the following scenario becomes apparent. ENERGY AND MASS FORMS DON'T JUST COME INTO EXISTENCE, LIKE IS SHOWN ON THE RECIPROCATING LINE GRAPH. The establishment of momentum into space requires all three dimensions in order to form 4-dimensional mass. The following diagrams describe how the initial transformation from zero space manipulates to form energy and mass. The onstart, from 0 to 1, moves like the knight piece does on a chess board (one position over and three positions down).

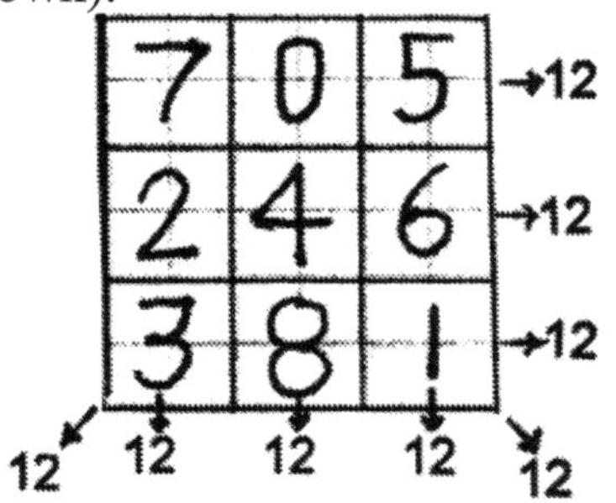

TOTALLY ISOTOPIC

Math scheme of the Pythagoras 3X3 set shown at the base of page 77.

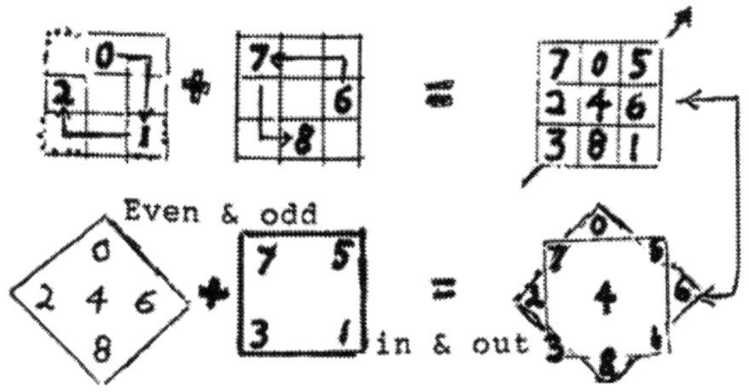

ZERO-POINT OF ENERGY UNIT

Notice of the line-graph, that the numbers are directly across from each other as they are on the math scheme on the 3X3 chart.

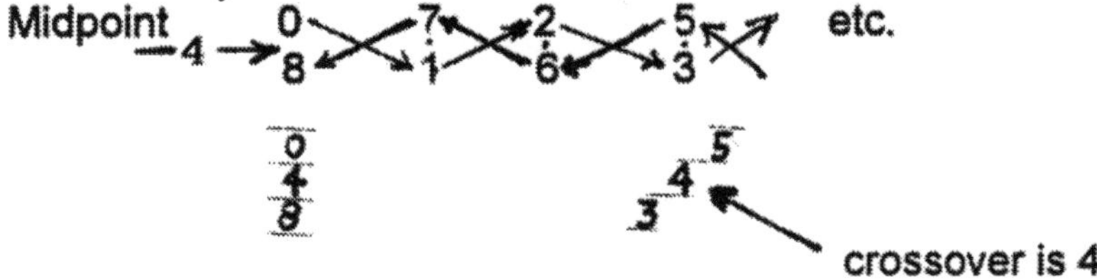

IF A SYSTEM IS TOTALLY ISOTROPIC, IT CAN MAINTAIN AN EXISTENCE FOR A WHILE IN SPACE. 12 RELATIVE TO THE VIEWER FROM ALL SIDES.

The problem, when constsructing the math locus on a squared positional sheet of graph paper, is that – *square positions* are gridlocked. There isn't any space between positions on the Pythagoras illustration of 3-4-5. In order to move from one position to another, it requires space.

To overcome this dilemma, maybe try to use circles as a substitute to replace the squares? Circles are okay, but don't properly solve the problem. What's needed is something that falls between a circle a square. The difference of the perimeter of a square compared to its diagonal, taken to its nearest fraction, is 20/7. The difference of the circumference of a circle compared to its diameter is 22/7. Therefore, let us try 21/7 which is 3. It lies between curve and straight. The following descriptions may help solve the problem.

Hilbert all evened space (gridlocked) *no room* to move	Curvature of Hilbert space like a rhombus

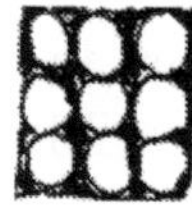

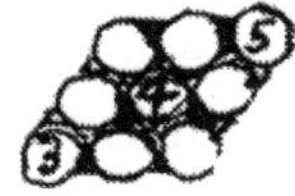

Circle-graph *has room* to move between the circles	Closed packed spacing still lets room to move

20/7 = 2.857142857142 ⟶ infinity 857 142 857 142 ⟶

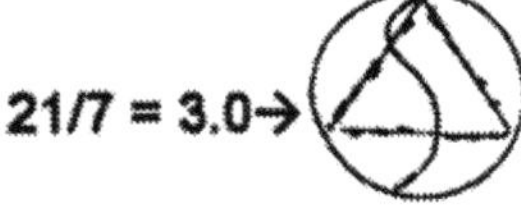

0 1 2	7 0 5	8 7 6
3 4 5	2 4 6	5 4 3
6 7 8	3 8 1	2 1 0
NO	**YES**	**NO**

22/7 = 3.142857142857 ⟶ infinity

$\overline{63/7}$ = 9

ONLY THE CENTER ISOTROPICAL 12 SYSTEM IS ABLE TO EXIST.

(SEE OCTONION ARITHMETIC DISPLAYED UPON pg 101. (explains 21/7)

142	857	142	857 ---
857	142	857	142 ---
999	999	999	999

maximum 9 [complete]

The smaller 20/7 square form system exactly fits in the 22/7 circular system. This means that by rotating the diagonal of the small

20/7 square form, it becomes "superpositioned" into the larger 22/7 circular system. 20/7 + 22/7

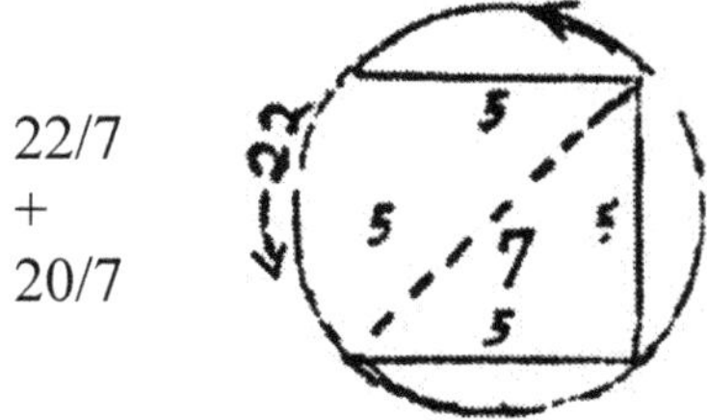

Both the square system and the circular system maintain the same length (7) for the diagonal and diameter of the systems.

This phenomenon is the onstart of natures endeavor to square the circle, so that each system contains equal areas, which is impossible to accomplish. Some might say, a cube form is able to be constructed so that it would hold exactly as much liquid as a sphere form. This may be correct to the drop, but liquid consist of drops --- drops consist of atoms --- atoms consist of subparts, which come together via the way of 20/7 which is an irrational infinite difference.

Now that the zero-point of energy is established as being the 3^2 (9 positional) system, it can be superpositioned into a Planck energy unit of energy.

4D space is activated into zero-point mode.

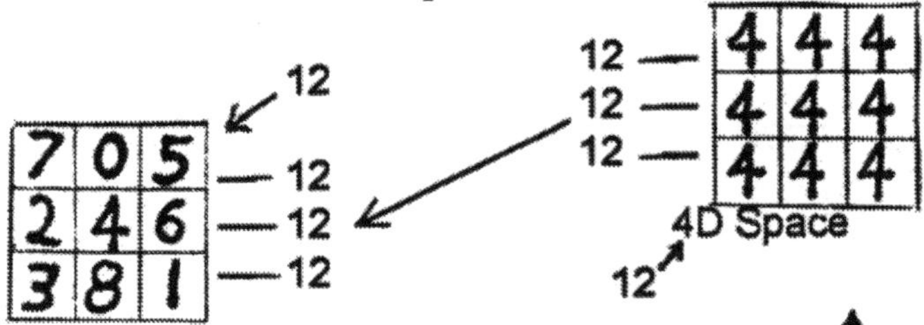

A graphed math scheme for Einstein 4th Dimension space-time constant, that's in accord with Hilbert's orthogonal (all evened) end-less space potential.

A symmetric tensor of the 9th rank, containing 81 scalar (even) and vector (odd) components within its system. This superpositions into a 9X9 graphed form having 40 *even* integers and 40 *odd* integers and 1 *even* 0 integer, which make up the math scheme of the momentum in the energy system. It is charted as:

The center (3X#) chart reverses as the momentum comes back *down* the chart.

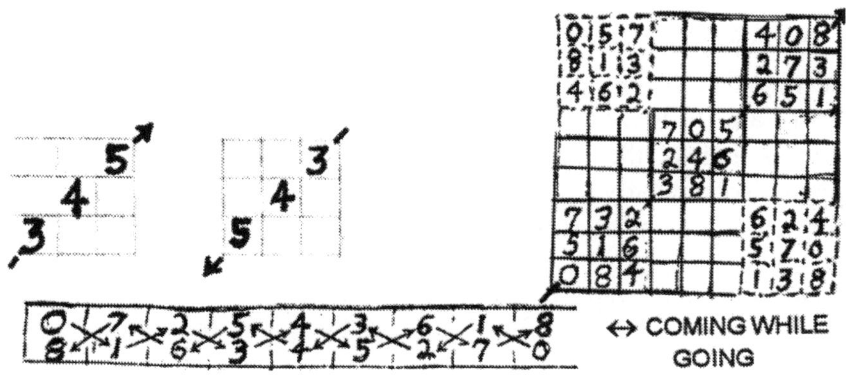

LINEGRAPH OF 3X3 (centered in 9X9 chart)

18 positions create two 9X9 charts.

Notice that the timing sequence of the momentum is coming/while/going. This causes a reciprocation which crosses back and forth (up and down) the diagonals creating a presence of energy within.

3-D superpositioning

00	01	02	03	04	05	06	07	08
09	10	11	12	13	14	15	16	17
18	19	20	21	22	23	24	25	26
27	28	29	30	31	32	33	34	35
36	37	38	39	40	41	42	43	44
45	46	47	48	49	50	51	52	53
54	55	56	57	58	59	60	61	62
63	64	65	66	67	68	69	70	71
72	73	74	75	76	77	78	79	80

The chart to the left is a natural consecutively occurring field in space. It rearranges to adhere to what is happening in its surrounding area. It is the potential for a Planck energy unit.

00 79 02 77 04 75 06 73 08

80 01 78 03 76 05 74 07 72

Partial line-graph
81 Positions

The superpositioned (3X3) at the center of the above chart cannot be arranged isotropic. Also, the energy move (1 over 3 down) starts at 00 to 11 and 11 to 22 and 22 to 33 and stops at 44. Then, restarts at 55 which is below the centered 3X3 chart. Superpositioning cannot happen isotropically causing the diagonals to be shifted as if reciprocating. Coming (up) – going (down) the diagonals. [See Planck chart].

The relativity aspect is aligned across the top row and down diagonally as:

01→02→03→04→05 etc.

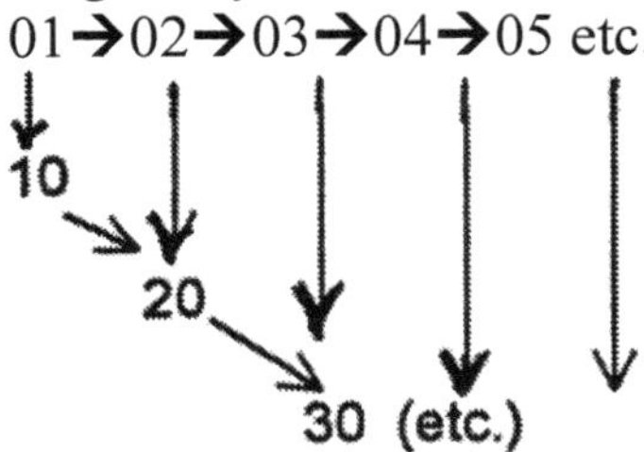

4-D superpositioning

00	47	30	49	32	15	62	17
29	59	45	02	61	18	13	34
46	01	48	31	14	33	16	63
51	28	03	44	19	60	35	12
04	43	24	55	08	39	21	59
27	52	07	40	23	56	11	36
42	05	54	25	38	09	58	21
53	26	47	06	57	22	37	10

The chart to the left depicts an isotropically arranged (energetic) field that peels off 15 positions, as the exchange of momentum, from a pi-meson to its adjacent segment of a neutron. Described on page 40.

The center (3X3) chart can't reverse.

The rows and columns each have the total momenta 252. The diagonals are arranged to *transfer* the 137 strong force. The left diagonal totals to 202, whereas the right diagonal totals as 2X137 (274). Isotropically enegetic means the 1 --- 3 transformation move contacts all 64 positions consecutively, as shown connected by lines on chart.

This printing omits many of the fundamental systems that play a part at the onstart structuring of mass forms from zero-mass. Sub-quark forms, quarks, lepton fields, numerous electron preactivation chartings etc. These are shown in the book called "*The Price Principles of the Heisenberg Zone*".

Photons contain momentum and energy in their system. The basic scheme of the fundamental, "zero-point", starter unit that superpositions into the Planck (h) unit is displayed page 28. (done in close-pack circles). The superpositioning can occur in different manners relative to which way the system is observed from the underlying area of the many aspects that tie together into the way zero-point units superposition, are explained as:

Mid

8 1 6 3 4 5 2 7 0 7

0 7 2 5 4 3 6 1 8 ←Mid 7.5

8

focused

Line graphs of the systems show a ½ difference at the line-graph midpoint area. This ½ causes a change of formation in systems.

LINE-GRAPHS RECIPROCATE

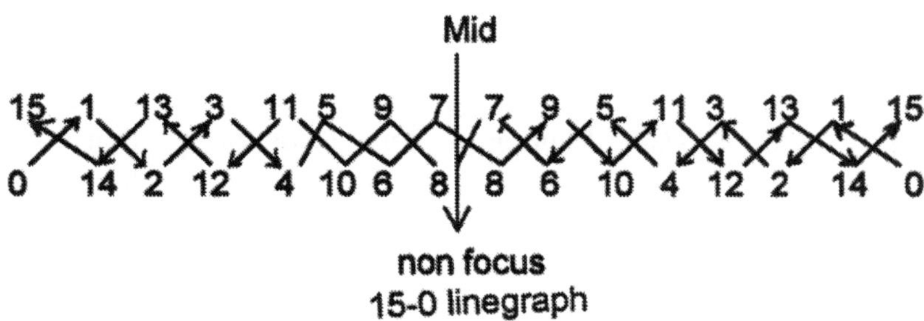

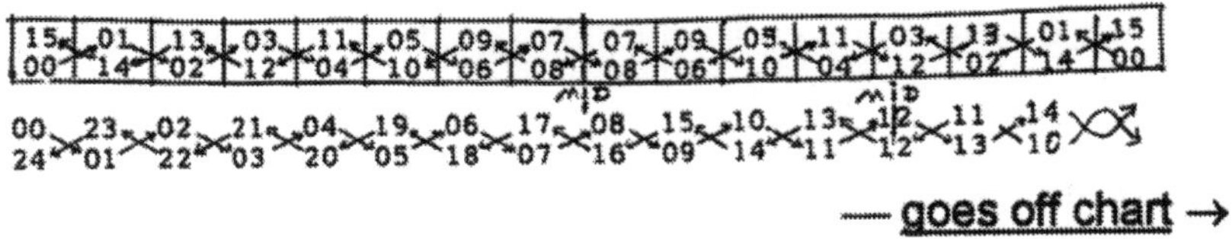

The first line-graph above depicts the 3X3 zero-point chart of a Planck energy system. The second depicts the 4X4 zero-point of an electron chart, to be described later. The 3rd depicts the 7X7 zero-point of a proton particle chart (also seen later). The 15-0 linegraph, displayed second, is invalid because the arrangement of math and geometry are equalized into position in systems that function from Hilbert space. The zeros can't be ignored, because they balance math with cosmic geometry. Example:

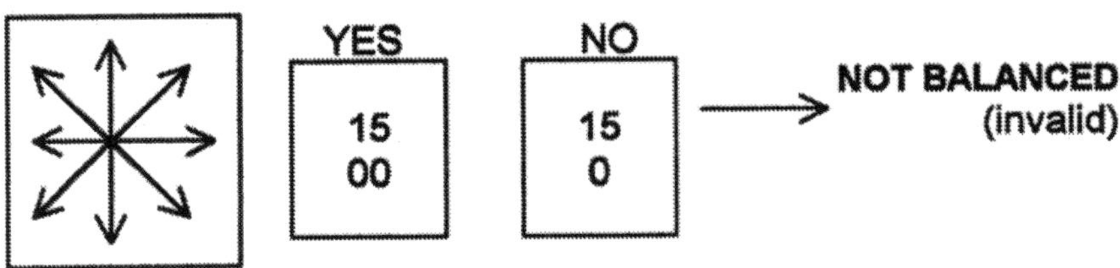

The result of correcting the invalid linegrah solves the long-sought question concerning the Reimann zeta function which goes as:

The LAW OF TRICHOTOMY states that, "every real number is either *positive (+) negative (-)* or zero. L.E.T. BROUWER, a Dutch mathematician, construed a mental image, where he imagined a number that is neither positive, negative or zero, that sets ½ to the right of where the zeta function equals 0 called Brouwers *counter-example* to the law of trichotomy.

BERNHARD RIEMANN another mathematician, worked out a scheme called, *The Riemann Zeta Function*, that begins $1+1/2_z+1/3_z+1/4_z+$---etc. sets the z into place as a "complex number", at which the zeta function equals zero. Riemann figured that these roots all contain real number parts that =1/2, they lie on a lien parallel to the imaginary axis and ½ unit to right.

Riemann's postulate was the most unique unsolved problem in mathematics until this writing illustrated the two-faced zero that balances geometry with math. *00=0 0=1/2* (onstart).

It is imperative that any explanation concerning the many structural aspects and the superpositioning schemes of subatomic systems be illustrated both *geometrically* and charted *mathematically*. Math and geometry are integrally related copartners that figure in the structure of mass forms. Nature tends to keep math and geometry together in unison.

HEISENBERG seemed to realize that momentum and position cannot be focused instantly in space at the same time. Why not?

NEWTON: Concerning speed --- charge *A* reacts immediately to charge *B*? Issac Newton's answer --- The act between A and B occurs instantaneously.

EINSTEIN'S rebuttal ----- "Nothing is faster than the speed of light" (186,274 mps).

I think that both Newton and Einstein were correct. My reasonsing is, "what is *instantaneous*"? Spaced position *A* reacts to spaced position *B*, which is already there! How fast is already there? [Instantaneous].

About ---- "*Nothing* is faster than the speed of light". What is nothing? Space is nothing. Therefore, space itself must be instantaneous! [When probed]. So, space being nothing, it is already there! Einstein did not seem to realize that space is nothing. Nothingless – already *here-there*-everywhere! It is absolutely impossible for the human mind to conceive the meaning of instantaneous, as such. Our brain waves need time in order to function. For the least action to exist, it requires time. Instantaneous requires *no* time, because its non-existence state is already there. The space-time idea, devised by Einstein, refers to *space* as being constant and *time* also as being a constant. [Until probed].

Space as being constant, anywhere and everywhere, engulfs everything. Therefore, when a disturbance occurs in space, the engulfed area endeavors to straighten back into the all evened constant space. Ata the onstart of a disturbance, such as happens when an electron "jump" its orbit and emits a photon, a 3-dimensional system structures that arranges into the zero-point unit of energy that becomes a photon. Photons carry energy and momentum, as well as angular momentum. The following chart illustrates the manner that the zero-point system is structured, from which photons and Planck energy units are derived. A disturbance in space, being totally engulfed, is an instant occurring in instantaneous space.

In order to assist the mind to grasp the manner that a least action can occur from an instantaneous state and set up fundamental subatomic systems, the following systems will be displayed in stages.

How things come into being from vacant space.

This paragraph briefly tells the most basic math that nature incorporates to transform space into the onstart of subatomic sized structures. The 3X3 system of the isotropically arranged Pythagorean 3-4-5 right triangle has rows and columns that total 012 each (12 as *we* sense it). I named 012 a "aristarter" in respect for Aristotles consecutive idea of how things proceed in growth. Another aristarter is 345 as found on the diagonal of the 3X3 system of Pythagoras 3-4-5 triangular scheme. Also to accommodate the 120 that presents itself as the total of the momenta found in the 4Dzero-point 4X4 system, I named 120 a "mixed aristarter". The reason for the same digits 012 and 120" as both being starting functions, is how the dimensions and proportions are observed relative to the cosmos (probed or viewed if you wish). Some examples are 102 (viewed from center 1st.) 102 and 201 are found on the chart of a proton particle, shown later. 021 is the total of momenta that is in the final 3X3 set, up the diagonal, of the Planck energy chart. The 1st 3X3 set starts with 012 up the diagonal of the Planck energy chart. I forget where 210 fits in but, I'll look it up in my charting files. 012-021-102-120-201 and 210 will be observed showing up on the superpositions of energy and mass charts later in t his article. Also 123-132-231---how else could the onstart of mass forms be charted?

ESTABLISHING AN ACTION THAT BRIELFY EXIST (Done in Stages)

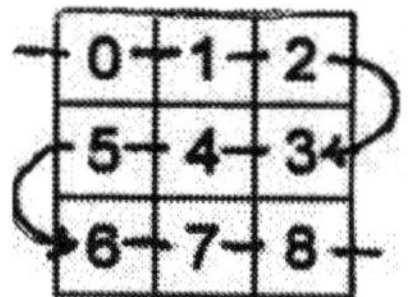

This chart illustrates a simple consecutive happening that occurs in 3-dimensional space

To better understand how sub-atomic particles come to exist in space, closely examine each chart [use scientific notation].

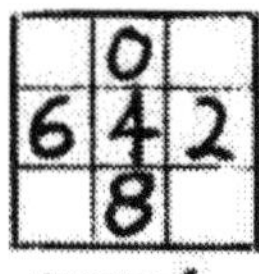

This all evenly mathed area, illustrates a segment of Hilbert orthogonal averaged out evenly space having no motion. In order for motion to occur, odd and even is a necessity.

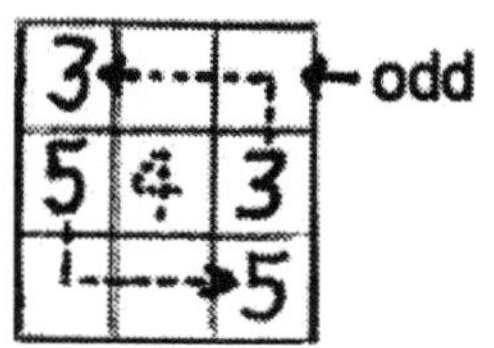

Nature employs a *1 position* over and *3 positions* down movement that brings odd amounts into the even Hilbert 3-dimensional space. Straight line locus

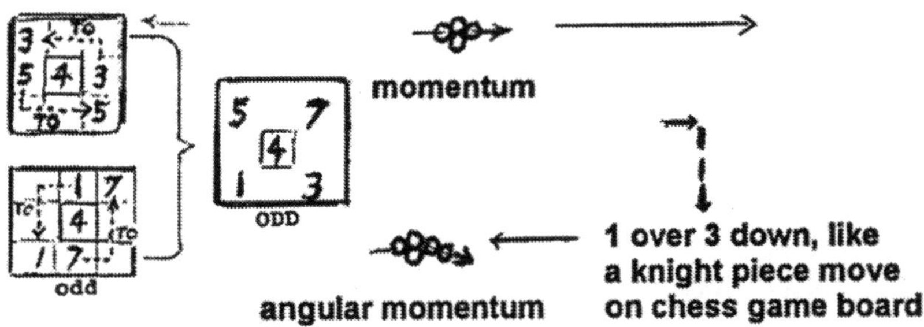

A gridlock charting of the previous actions that complete the space transformation.

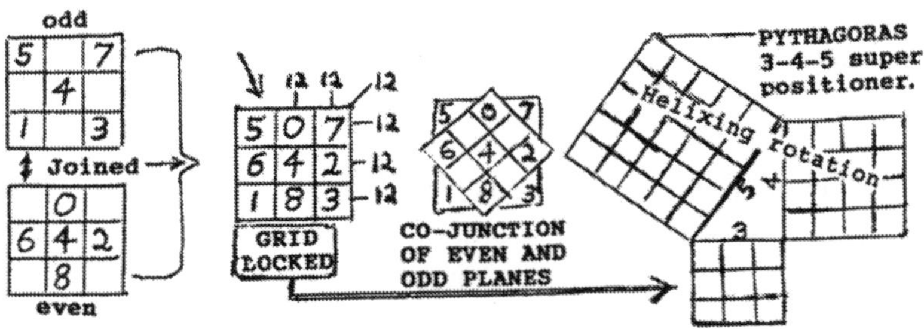

THE GRIDLOCK CHART (ABOVE) IS AN ISOTROPIC SYSTEM --- EACH ROW = 12 AND EACH COLUMN = 12 DIAGONALS ALSO = 12. The isotropoic aspect enables the system to maintain a brief time in instantaneous space.

The gridlocked zero-point least action (3X3) starter is displayed above. It is the super-positioning base involved in the Pythagoras 3-4-5 right triangle.

Space, being considered as constant, is instantaneous in any given area, such as a position in space, because, it is already there. Energy is required to *move* between positions in order to overcome the instantaneous (stilled) state of the zero-space positions. The requirement of energy to overcome instantaneous, is somewhat like moving a block of concrete from its stilled position. Virtual momentum (C^2) functions at the fundamental area of zero-space due the instantaneous state of zero-space being constant. [I hope you are not one of those "HARD TO MAKE SEE" people].

Some subatomic researchers refer to a single position in space as a graviton. During the event when a potential positioned area in zero-space is "disrupted", it reverts into two rhombic-like planes. The result of the disruption of the instantaneous state of Hilbert orthogonal space is a fluctuating curvature of the two converging planes (even – odd). A fluctuating system emits an

"outward" *going* force ↖ ↗ of virtual momentum (at light speed)
↙ ↘
away from the instantaneous

space system and conversely, drawn in a "inward *coming* toward force action. The system is continuously gravitating. ↘ ↙
↗ ↖

The foremost basic tunneling action occurs mathematically drawn as 01 ←→ 10. This is the way that instantaneous space, when disrupted, observes the action (relative to the *observer* which in this case is zero-space itself).

↘ ↙
The reason why gravity is pulled "inward", ↗ ↖ is because the isotropically arranged gravition-like system begins at 3 dimensions in space, as depicted in the following chartings.

Difference

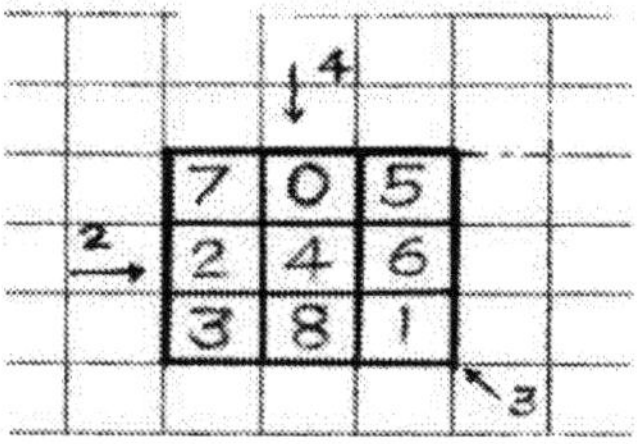

7	0	5
2	4	6
3	8	1

INWARD
The total momentum
in the system is 36

ISOTROPICALLY ARRANGED MOMENTUM rows – columns & diagonals each = 012

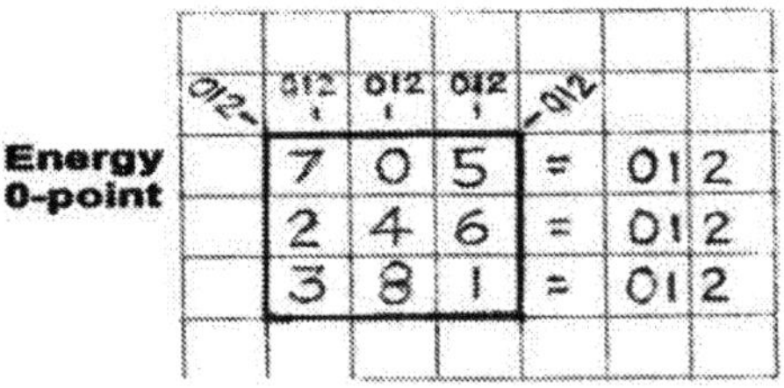

7	0	5	=	012
2	4	6	=	012
3	8	1	=	012

Being balanced at 012
makes the system able
to exist in space

012 "activates" the
instantaneous area
into a zero-point
energy system.

From instantaneous
into energy occurs
at light speed.

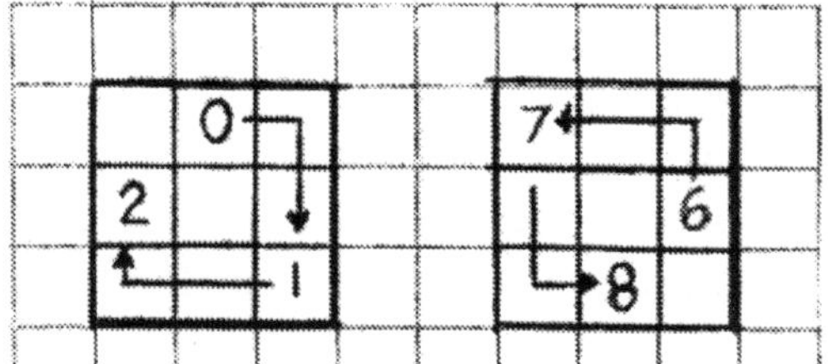

Angular momentum
binds unit intact

straight line and
angular momentum
done in circles

Momentum and also angular momentum
are structured in all of the atoms

Hilbert orthogonal space, being all evened, is situated in space having the potential to accept *odd* amounts into its orthogonal averaged out *evenness*. The 2-dimensional math gridlocked charting of the virtual momentum in a consecutive grouping, that signifies 20/7 in con-junction with 22/7, sets up as:

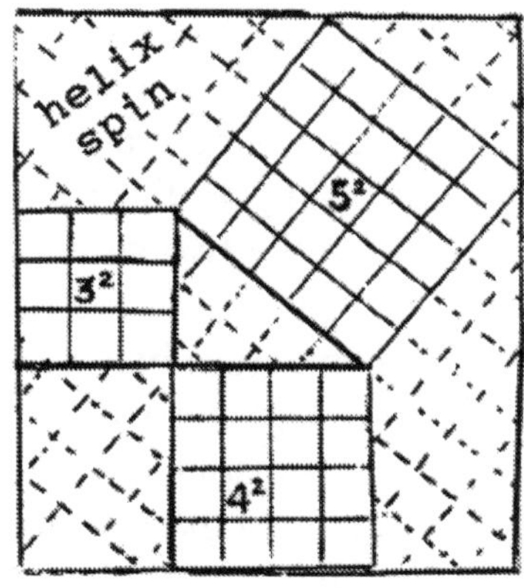

This is the square of 3 that is depicted on the Pythagorean 3-4-5 right triangle scheme that joins with the base (4^2) to form into a mass-energy relationship. The*zero-base* of superpositioning begins with the Pythagoras right triangle scheme.

For one to grasp the basic significance of superpositioning, like as shown on the Pythagoras 3-4-5 right angle scheme, examine the 4^2 zero-point chart and note that, as the perimeter of the 4X4 system expands into larger size systems because of excess energy being absorbed through the midpoint, that the diagonals at the 4 corners of the 4^2 system keep maintaining a *constant* momenta total that carries from the 4 midpoint positions outward indefinitely. Study the following example.

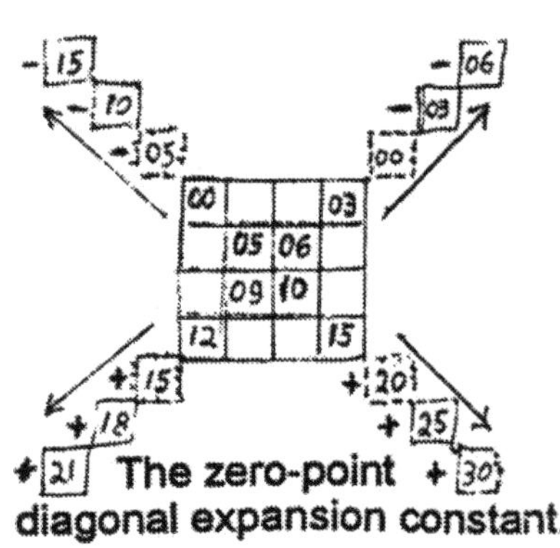

The zero-point diagonal expansion constant.

-05 -03→
00 +03→
+05+06 } 5+6+9+10=30 }: 30} =30
+09+10
+12 +15→
+15 +20→

Continuing the left diagonal by 05 and continuing the right diagonal by 03 will hold the diagonal corners to total 30 even if the system is superpositioned infinitely.

The "Layout" of a zero-point energy unit, displayed next, superpositions into mass/energy relationships. One product of this superpositioning, when multiplied to the capacity of enough energy able to structure life forms, is the formation of a helix shaped (360 degree) segment of a DNA molecule. It is superpositioned from the total momenta (036) that is in a zeropoint energy unit. The chart details the inner workings of the momentum that crosses the helical field between the 2 phosphate strands. The engery zero-point is superpositioned in a manner that crosses from strand to strand at 036 degree intervals *arranging* atoms into aminos according to the field capacity. Ten sets of amino acids are rotated between the two strands which completes the 360 degree circle-squaring endeavor. Instead of pi taken to its nearest fraction (20/7) true pi is drawn across between the strands which is capable of creating *all* of the characteristics of life forms. This huge illustration of a DNA molecule is displayed on a chart that is supplied along with the book, *The Price Principles of the Heisenberg Zone*. True pi, taken to only 70 decimal places of its circle-squaring endeavor, tightens as:

π=3. 1415926535 8979323846 2643383279 5028841971 6939937510 5820974944 5923078164

MOTION CONTAINS MOMENTUM

It seems that ever since the apple fell down toward earth from the tree, we have attempted to reason why. Sir Isaac Newton – 1687 – introduced his universal law of gravitation which deals mass, force and acceleration. F=Ma. He also formulated the acceleration difference of falling bodies. His inverse square of the distance law coincided with the aspects of electric force. This type of data does not reveal what happens to the motion itself. Motion is the thing that must be explained to solve the gravity scenario. So, motion moves toward stillness. This book has revealed the *action* coming/while/going that is structured in mass particles. Earth, meaning where the apple hits ground, is a hard (stilled) place that stops the inertial apple. Motion *within* the sub-atomic particles reversed from coming (like negative force), to going (like positive force). See page 67. Consider the moon in its gravitational relationship with earth. It creates gravitating motion that causes the ocean waves. The water in the ocean is not hard grid-locked like the terra firma of earth. Therefore, the water does not still the motion until it penetrates enough distance into the water. This penetration causes the waves and tides and even has an effect on the winds. You would think that space is still, so why doesn't mass go up into space stillness, instead of down to earth? Mass possess inertia forces within its structure that manipulate according to the inverse square law. If it is far enough from the surface of earth, it will float in vacant space. Now, consider the sun. It is located at the centroid of our solar system. That is where solar gravity motion hits bottom, so to speak. At the wee small quantum area of the gridlocked protons (sets of 4 connected protons) called protonion give off the positive like *going* away from motion. Light etc. Now you should better understand the riddle of gravitation and quantum gravity. **Again see data on page 75**.

The following diagram may be used to better grasp the electrical field when it undergoes the coming/while/going current system. This field is also shown on the DNA starter chart displayed elsewhere in the *Heisenberg Zone*

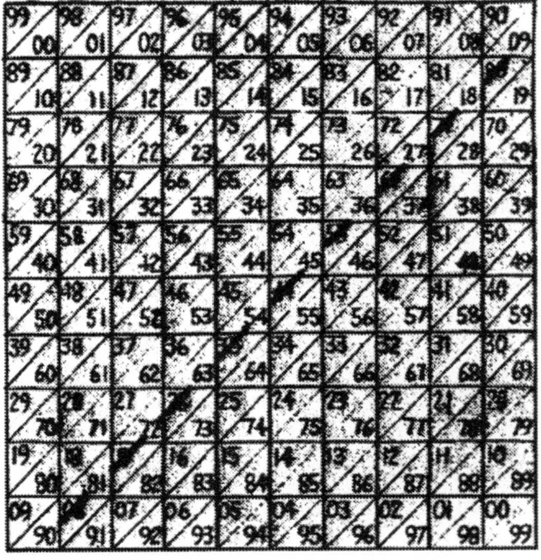

99 00	98 01	97 02	96 03	95 04	94 05	93 06	92 07	91 08	90 09
89 10	88 11	87 12	86 13	85 14	84 15	83 16	82 17	81 18	[illegible] 19
79 20	78 21	77 22	76 23	75 24	74 25	73 26	72 27	[illegible] 28	70 29
69 30	68 31	67 32	66 33	65 34	64 35	63 36	[illegible] 37	61 38	60 39
59 40	58 41	57 12	56 43	55 44	54 45	[illegible] 46	52 47	51 [illegible]	50 49
49 50	48 51	47 58	46 53	45 54	44 55	43 56	42 57	41 58	40 59
39 60	38 61	37 62	36 63	[illegible] 64	34 65	33 66	32 67	31 68	30 69
29 70	28 71	27 72	[illegible] 73	25 74	24 75	23 76	22 77	21 78	20 79
19 80	18 81	[illegible] 82	16 83	15 84	14 85	13 86	12 87	11 88	10 89
09 90	[illegible] 91	07 92	06 93	05 94	04 95	03 96	02 97	01 98	00 99

← TOP CHART IS ISOTROPIC DUE TO BEING STRUCTURED WITH ALL SQUARES.

THE LOWER CHAET IS *NOT* ISOTROPIC DUE TO BEING CONSTRUCTED WITH ALL RECTANTLES. NOTE THE DIAGONAL CURRENT SHOWN IN THE TOP CHART. IT IS ARRANGED IN A STRAIGHT LINE AS: 9-18-27-36-etc.

THE IDAGONAL CURRENT ON THE BOTTOM CHART *STAGGERS.*

ODD & even momentum shifts in space on two planes.
Deviation from isotropic causes charge to occur in a system.

00 99	98 01	02 97	96 03	04 95	94 05	06 93	92 07	08 91	90 09
10 89	88 11	12 87	86 13	14 85	84 15	16 83	82 17	18 81	80 19
20 79	78 21	22 77	76 23	24 75	74 25	26 73	72 27	28 71	70 29
30 69	68 31	32 67	66 33	34 65	64 35	36 63	62 37	38 61	60 39
40 59	58 41	42 57	56 43	44 55	54 45	46 53	52 47	48 51	50 49
50 49	48 51	52 47	46 53	54 45	44 55	56 43	42 47	58 41	40 59
60 39	38 61	62 37	36 63	64 35	34 65	66 33	32 67	68 31	30 69
70 29	28 71	72 27	26 73	74 25	24 75	76 23	22 77	78 21	20 79
80 19	18 81	82 17	16 83	84 15	14 85	86 13	12 87	88 11	10 89
90 09	08 91	92 07	06 93	94 05	04 95	96 03	02 97	98 01	00 99

Curvature added to a flat non-isotropic chart causes *deviation* from straight line current to occur. see page 88 and also the plate displayed on page 80.

(This caption was placed below the picture in the original.)

CONCERNING THE CAP CHART

The CAP Chart is larger than a map size drawing. Cap means atomic capacity. The atoms are arranged on the chart starting with hydrogen and ending at neon (10 atoms). The chart has 10,000 positions set on a 100 X 100 field area. The positive field capacity sets the atoms into place at the square of the distance from each other. See the partial cap chart displayed on the front page of this book and note that the current goes up the diagonal, then across the top and returns back down the opposite diagonal. It crosses at the midpoint, where the centroid of the nucleus is located. This is similar to the sun being the centroid of the solar system. Recently I read that Heisenberg constructed a chart that arranged atoms at the square of the distance. I wonder if his charting is similar to my Cap chart? Sir Arthur Eddington made at 16 X 16 chart intended to solve the 137 force scenario. At that time the strong nuclear force was thought to be 136 times stronger than electromagnetic. When 137 was determined as correct, he was ridiculed, but when a 16 X 16 graph is dissected diagonally, the result picks up exactly one additional position, making the 137. See Ben Franklin's 16 X 16 chart on page 25. Being that the CAP Chart is too large to be presented in this book, we are contemplating having the chart printed along with the much larger DNA chart which measures about 4 foot by 6 foot. A helix field. This is necessary because a large all inclusive CAP chart could not be provided by the publisher. We placed a small partial segment of the CAP chart in the frontice piece of this book for your reference. The partial CAP chart is limited in scope since it only illustrates the positive field area that houses a hydrogen atom and a helium atom. The additional atomic data is displayed on the larger complete CAP chart. Both charts should be readied for mailing by the time this book is in your hands.

Please mail all request and inquiries to:

PENNICK SCIENTIFIC ENTERPRISES, P.O. BOX #*@ CARNEGIE PA 15106-0382, or consult *pennscience.com.* http://www.pennickscientificenterprises.com/about.html

Hilbert space definitions

Rotational symmetry – by turning a square ¼ or ½ turn, it remains unchanged relative to the viewer.

Reflection symmetry – top ½ of a square mirrors the bottom ½.

Translational symmetry – the eveness of all squared Hilbert background allows the square to be shifted up, down, or sideways & would blend exactly into the background relative to the viewer.

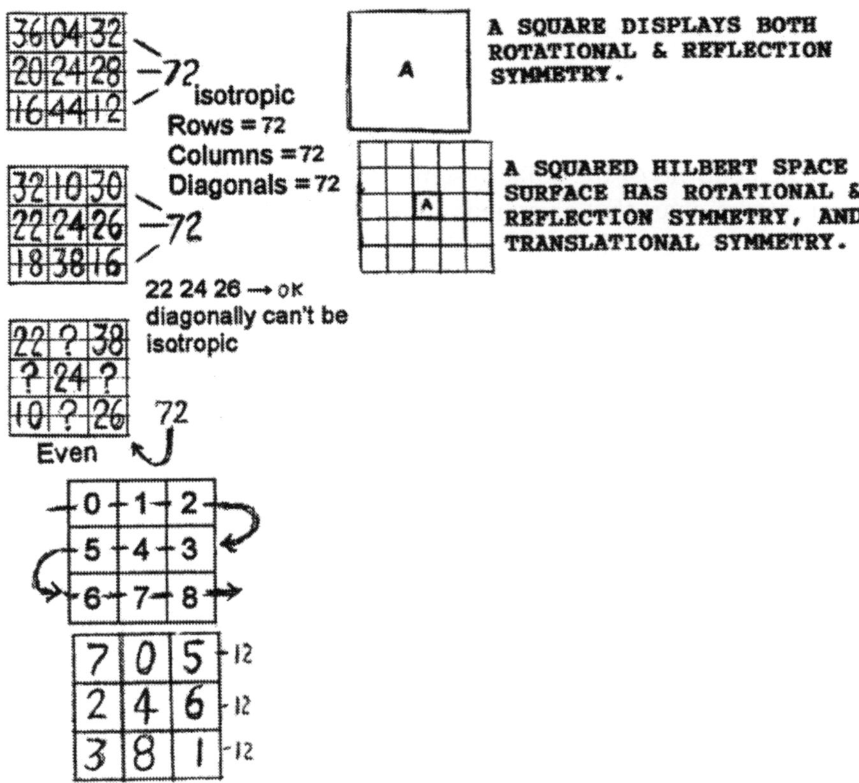

The eye view of the 2-dimensional charts that are displayed in this book shows the momenta in the system as if being girdlocked togther on a flat surface. The "Layout" shown next illustrates the space-time, even and odd position transformation of the momenta that is depicted on the chart, as being arranged in 3-dimensions.

0	0	0	0	0	0	0	0	0
1	1	1	1	1	1	1	1	1
2	2	2	2	2	2	2	2	2
3	3	3	3	3	3	3	3	3
4	4	4	4	4	4	4	4	4
5	5	5	5	5	5	5	5	5
6	6	6	6	6	6	6	6	6
7	7	7	7	7	7	7	7	7
8	8	8	8	8	8	8	8	8

This 9X9 positional chart, drawn on flat 2-dimensional paper, shows the *overall* scheme of the area that is involved in the space-transformation that arranges into an isotropical system see page 67.

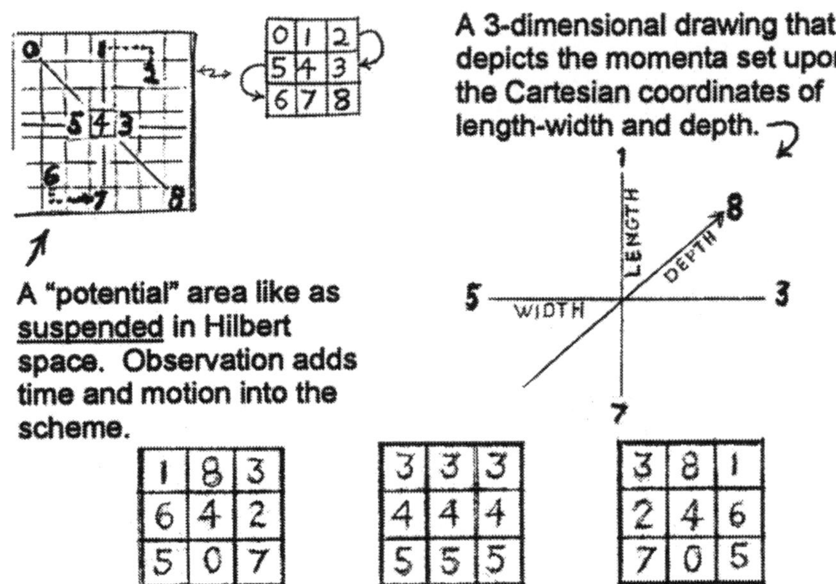

➚ When this stilled form chart sets into motion, the 3-4-5 aligns according to the change of the rotational direciton determined by the angular momentum timing sequence.

Now that the primary 3X3 zero-point of an energy unit has been established, a few of the many energy-mass systems that are superpositioned from zero-point systems can be charted.

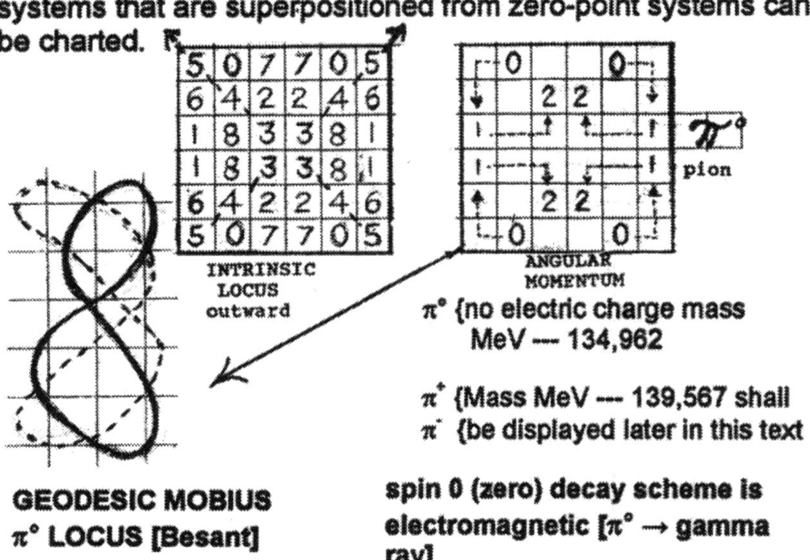

3	35	28	12	17	10	-105
29	4	33	11	13	15	-105
7	27	32	16	9	14	-105
30	8	1	21	26	19	-105
2	31	6	20	22	24	-105
34	0	5	25	18	23	-105

105 105

3	35	28	66
29	4	33	66
7	27	32	66
39	66	66	39

12	17	10	39
11	13	15	39
16	9	14	39
39	39	39	39

39

30	8	1	39		21	26	19	66
2	31	6	39		20	22	24	66
34	0	5	39		25	18	23	66
66	39	12	66	66	66	66	66	66

The isotropic 105 chart is the zero point of a π° unit

3					10
	4			13	
		32	16		
		1	21		
	31			22	
34					23

70-CORNER EXPANSION
***constant**

3 + 10 34 23 = 70
4 + 13 31 22 = 70
32 + 16 1 21 = 70

The mass of a pion π° is near 1/7 the mass of a proton particle.

BRIEF LIFETIME IS: 0.83 X 10-16 seconds.

The π° (pion units) assist in the transferring of all of the characteristics that form together across between the amino acid molecules of DNA. Characteristics stem form 70 decimal places of pi. See the decimal expansion of pi, taken to 70 decimal places (displayed page 50).

Now, that the *energy* zero point 3^2 [3X3 system] taken from the Pythagoras right triangle scheme has been illustrated, the next step is to decipher the zero-point of *mass*.

The zero-point of mass, from which more massive forms can be superpositioned, is 4^2 [4X4 system on the Pythagoras scheme]. The Pythagoras right-triangle scheme is "natures" basic action that superpositions into larger mass forms. From zero-space completely on up to the entire solar system.

00	14	13	03	=	30	00	27.5	22	16.5	66	00	41	31	30	=	102
11	05	06	08	=	30	11	27.5	33	44	115.5	11	50	60	80	=	201
07	09	10	04	=	30	38.5	49.5	5.5	22	115.5	70	90	61	40	=	201
12	02	01	15	=	30	16.5	11	5.5	33	66	21	20	10	51	=	102

-negative	zero-point Midpoint	+positive
zero-pointof	*15 atoms*	*zero-point*
Is Able to Expand (012)	Potentially sets up the atoms Onto CAP Chart	Binds into Proton
Isotropic	Semi-Isotropic	Staggered Isotropic
row – cols – diagonals	row 1 66	rows – cols – diagonals
add to 30	row 2 115.5	adds up to 201 ←→ 102
(4X30=120)	row 3 115.5	total ←→ 606
EXPANDS INTO	row 4 <u>66</u>	BINDS TOGETHER
137 ELECTRON	total 363.0→	

See Cap chart (atoms are set into place in order up diagonal) Hydrogen-5.5 Helium-11 Lithium-16.5 Beryllium-22 Boron-27.5 Carbon-33 Nitrogen-38.5 Oxygen-44 Fluoride-49.5. When these number amounts are multiplied by 99 (electron field) they are no longer potential amounts and become atoms [44X99=4356] Oxygen.

NOTE: 012-021-102-120-201 and 210 are all fundamental starters that begin the structure of systems at zero-space. [Relative to from which direction they are observed in zero-space]. The midway zero-point chart sets up the structure of atoms on the CAP chart by the use of advanced starter groups as: 123-132-213-231-312 and 321, arranged relative to the observer.

1/3 quart = 33 2/3 quark = 66 2 X 66 = 132 2 X 155.5 = 231

Momenta in a 100X100 electric field totals 99 (33+66). 132-231. The above shown zero-point of negative and positive systgems and also the mirrored observation results of 132 231 etc. play a large part in the structure of matter and anti-matter. The end result is determined by which way the probe is directed at the very onstart of observation. [Coming and going "*mirrors*"]. Now on with the show!

This proof column is to be used as a reference to decipher the underlying math and geometric functions that are captured by the charts that designate subatomic particles etc. that are displayed in this

CAP Chart article. The various charts superposition from the zero-point charts, up to the entire periodic table of elements.

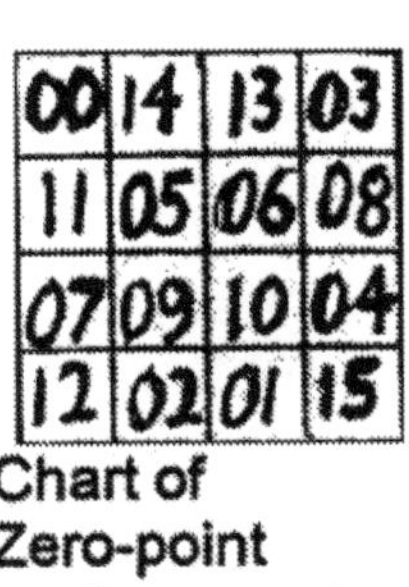

Chart of Zero-point

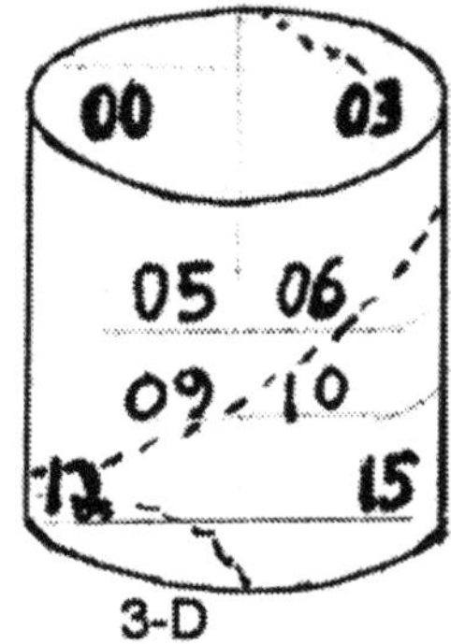

3-D

A 3-D view can be observed by placing the numbers on to a cellophane cylinder. The numbers show through same like drawn on the chart. Also, the crossover helix becomes clear.

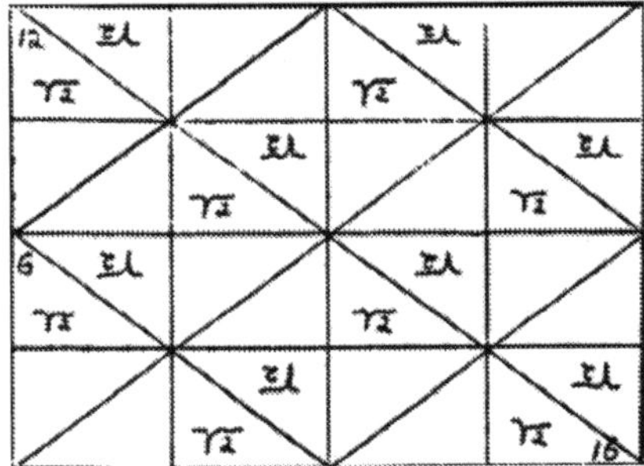

The multiple 2 left scheme displays how the Pythagoras 3-4-5 superpositions into larger formations as like the diagonal 6-8 and 12-16 rt. Triangle repositions on chart.

INTEGRAL PART OF SPACE

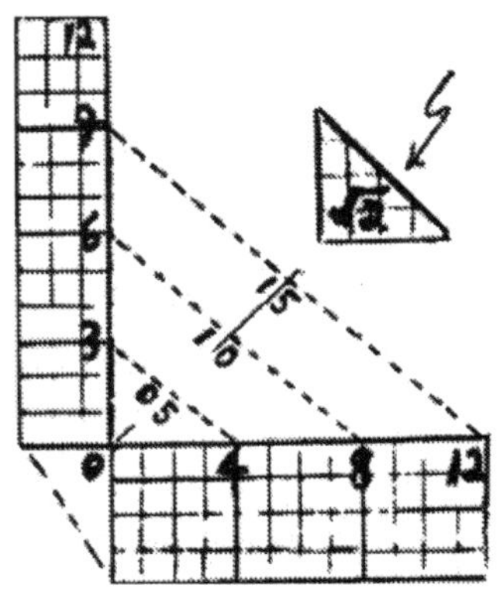

This graphed design shows how the √2 superpositions into mass forms, at *least action*, from 0 space, Energy (3X3 verticle 3-6-9 –12), activates the 0-05-10- 15 into a helically connected zero-point scheme, as displayed on charts illustrated above.

The above shown graph of the 3-4-5 Rt. Pythagorean scheme depicts expansion. It portrays the manner of the √2 is superpositioned, up from zero space, expanded by energy to a 15 positional hypotenuse. This creates the zero-point system that forms mass (see the 4X4 zero-point chart displayed above). The vertical arranged graphed section represents the 3X3 energy zero-point systems that function with the 4X4 graphed base, which is comprised of the evened Hilbert space background. The *entire scheme* should be displayed on a Hilbert space *graphed* background. I omitted the graphing on the dotted hypotenuse section in order to maintain clarity. The 4X4 base row has been displayed on a graph background to illustrate where and how the energy (3X3-off) functions with the 4X4 (even) creating a least action $\sqrt{2}$ system and super positions to zero-point chart and on up to the capacity where atoms form.

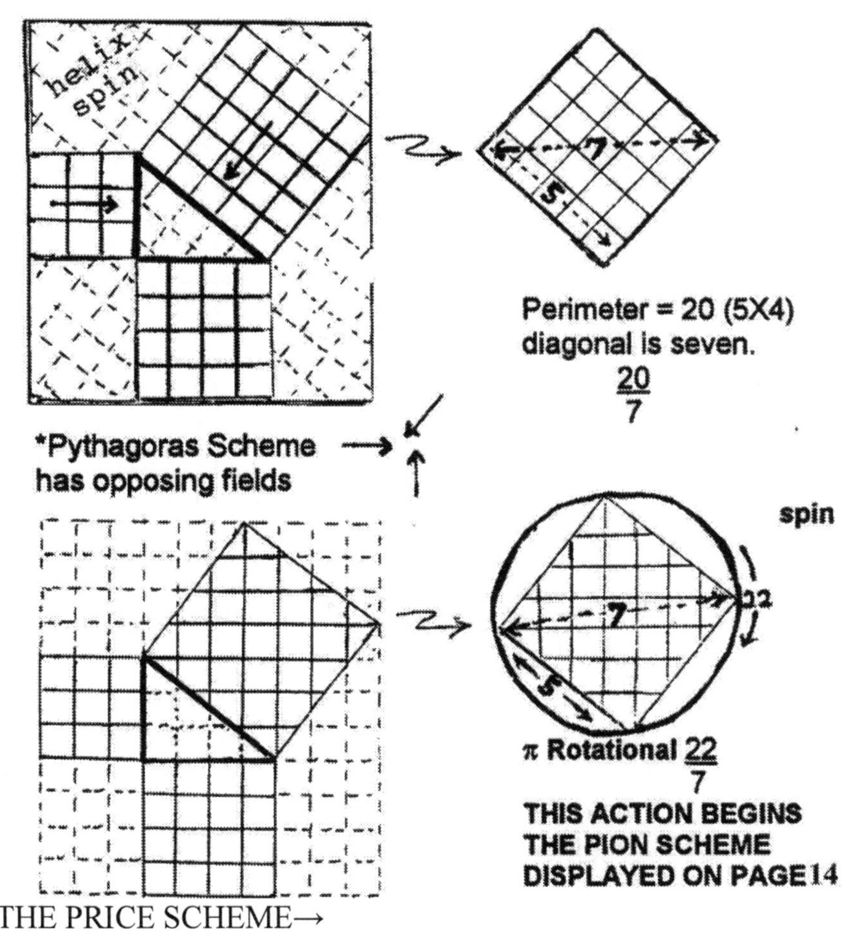

*THE PRICE SCHEME→
→

is in unified field→

ALL MASS EXISTS IN FIELDS --- OPPOSING FIELDS CAUSE ROTATION

EINSTEIN REALIZED THAT ALL MASS EXISTS IN FIELDS – HE TRIED TO CONSTRUCT A THEORY THAT UNIFIED THE FIELDS UNTIL HE DIED. Now that the most fundamental basic starting systems have been presented, the structural details of some major atomic particles can be illustrated. Sub-quarks, muon, kaon, etc. are set in the book "*Price Principles of the Heisenberg Zone*".

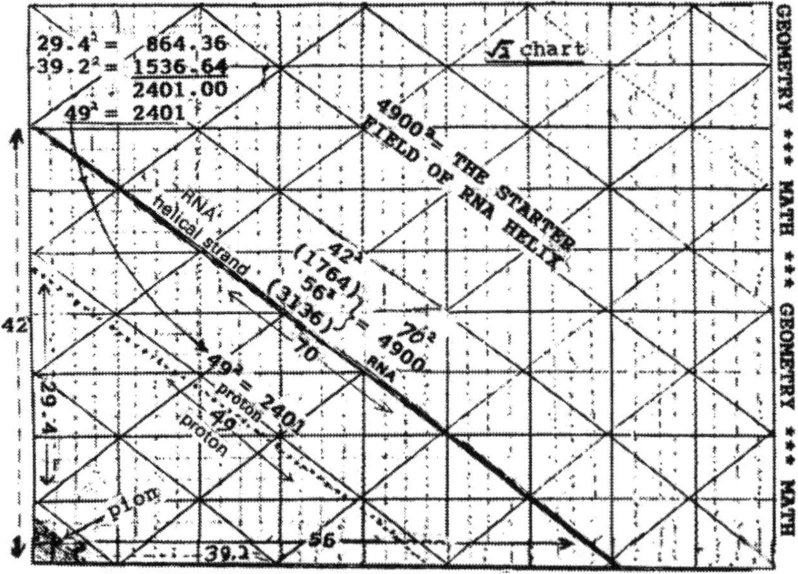

An illustration of the potential $\sqrt{2}$ field expansion scheme in space, from which particles of mass can be superpositioned. [This $\sqrt{2}$ field would be *small* beyond the range of electron microscopes – if it could distinguished from our imagination. The comparison of the dimensions of the size between a pion and proton is marked by the dotted lines on the $\sqrt{2}$ field. The *small* dotted right triangle scheme ★ seen at the lower left corner of the field, represents the zero-point scheme of a pi-meson (pion). The larger dotted hypotenuse (about 1/6 of the field) represents the zero-piont scheme of a proton. [The scale of each square block on the graph equals two].

Notice on the large $\sqrt{2}$ chart shown on page 90, there are 7 segments along the hypotenuse (70). Also 7 segments on the height (42) and 7 segments across the base (56). This isotropically arranged scheme develops into the *stable proton.* The *42* and *56* squared sets *generate* the 70 squared set into isotropic stability.

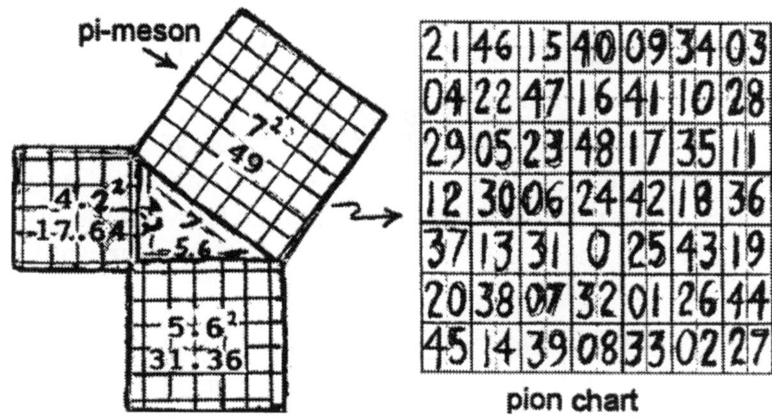

THE TRIANGLE SHOWN ★ AT THE LOWER CORNER ON THE BIG $\sqrt{2}$ CHART *Pg. 90* IS THE BASE FROM WHERE PI-MESON SUPERPOSITIONS.

A proton superposition via the pion scheme as: 4.2 to 42, 5.6 to 56. 07 to 70 as seen on $\sqrt{2}$ chart. [Illustrated later]

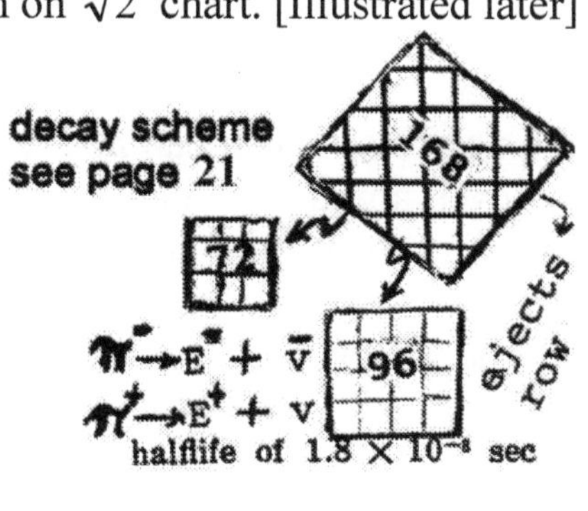

$\pi^- \rightarrow E^- + \bar{v}$
$\pi^+ \rightarrow E^+ + v$
halflife of 1.8 X 10^{-8} sec.

21	46	15	40	9	34	3
4	22	47	16	41	10	28
29	5	23	48	17	35	11
12	30	6	24	42	18	36
37	13	31	0	25	43	19
20	38	7	32	1	26	44
45	14	39	8	33	2	27

pi-meson is totally isotropically constant. note that all lines end at two #'s adding to 48.

any direction

21	22	23	6	5	4
27	10	17	42	43	44
3	26	31	48	35	28
45	38	25	0	13	20
96	96	96	96	96	96

46	15	40	47	16	29
2	33	8	1	18	19
34	9	12	41	32	11
14	39	36	7	30	37
96	96	96	96	96	96

Diagonal expansion holds constant at 96 (corner totals)

Trace where the numbers fall on the chart the resulting pattern always is 96. Later in this article the exchange between the pi-meson and its field will show the 96 group "cast off" in pi-mesons decay mode.

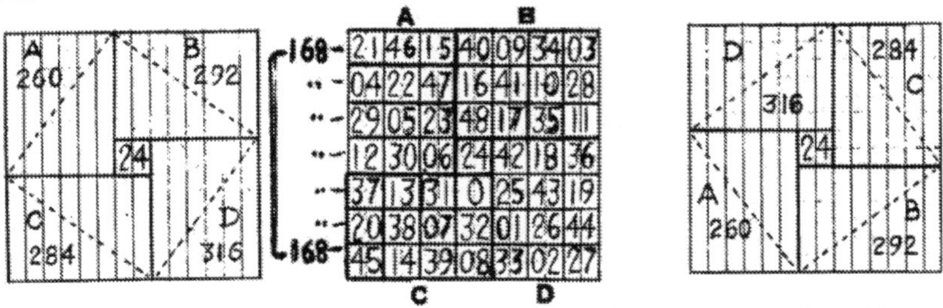

21	46	15	40	09	34	03
04	22	47	16	41	10	28
29	05	23	48	17	35	11
12	30	06	24	42	18	36
37	13	31	0	25	43	19
20	38	07	32	01	26	44
45	14	39	08	33	02	27

A pi-meson, being *totally* isotropically structured, allows an immediate exchange of the momenta of the strong forces in the nucleus of an atom occurring between protons and neutrons. Also, during the weak force interaction in the decay process of pions, the even and odd amounts of momenta separate. The odd amounts are ejected together into a neutrino, which can not be detected because it requires both odd and even amounts to detect a frequency within the evened Hilbert space. The even amounts revert into a all enened system as is displayed in the above portrayed pion decay process. [Also see page 21]. The following diagrams depict the symmetric decay of a pion.

THIS FINAL SECTION EXHIBITS THE MATHEMATICAL AND GEOMETRICAL STRUCTURAL DESIGNS THAT PROVE HOW SUB-ATOMIC PARTICLES ARE FORMED INTO

ENERGY-MASS SYSTEMS AND ALIGN ONTO THE CAP CHART.

By bypassing the smaller sub-quarks, momentum and angular momentum units, the next step up (in space systems) concerns *The Hamiltonian Principle of Least Action*, in reference to energy-mass units. The zero-point of energy is formed in 3-dimensional space having length-width-depth. This 3-D system is motivated by the angular momentum, which motivates the energy unit to develop into a 4-dimensionally "timed" unit.

The fundamental (zero-point) unit of energy is displayed mathematically in the charts that are displayed next. The 3X3 charts, shown below, can be traced way back to ancient China. The nine positional (3X3) chart is called a "Lo Shu Square". It is considered as a key to scientific wisdom.

The 3X3 square patterned, chart is also akin to the ancient Hebrew symbol of the Sigil of Saturn, which displays the angular momentum locus of the zero-point (3X3) unit. The unit is the zero-point of Planck's fundamental energy unit (h). The difference between the Planck unit compared to the Sigil of Satum and Lo Shu charts is that the Planck unit begins with zero, whereas the Sigil and Lo Shu are superpositioned up starting at 1 instead of 0. A few of the basic proofs are:

Zero-point space
base of Sigil

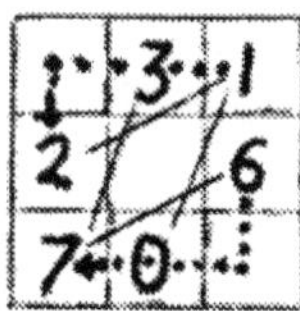

CONNECTS TO
SPACE AT 0

Zero-point of
energy unit

7	0	5	→12
2	4	6	→12
3	8	1	→12

ISOTROPICALLY BALANCED
AT 12 – CHART TOTALS 36

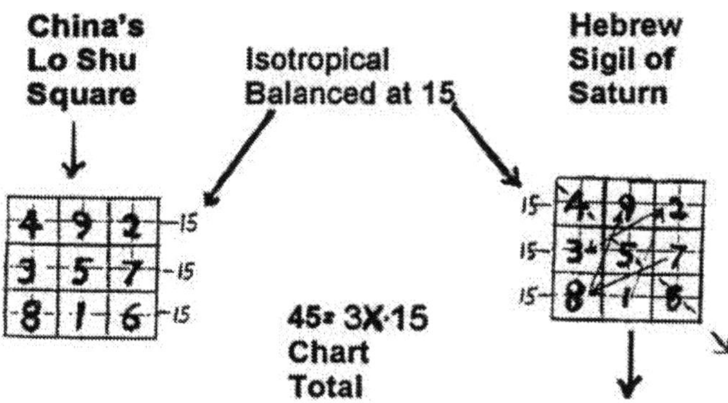

THE LO SHU AND SIGN OF SATURN BEGIN AT ONE and contain no zero. The zero-point of energy square is attached into space at zero. It superpositions into a Lo Shu.

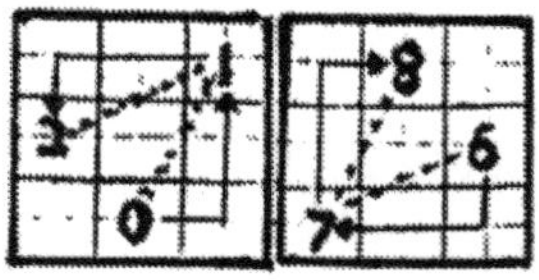

The solid lines between 0 & 1 and 1 & 2 depict the 1 over, 3 down angular momentum motion. The pattern of dot marks on a turtle's shell mimicks the numbers on the Lo Shu square. It is derived from its DNA scheme.

The 1 over 3 down move (like chess knight piece moves) hits on the same numerals that the sign of Saturn do.

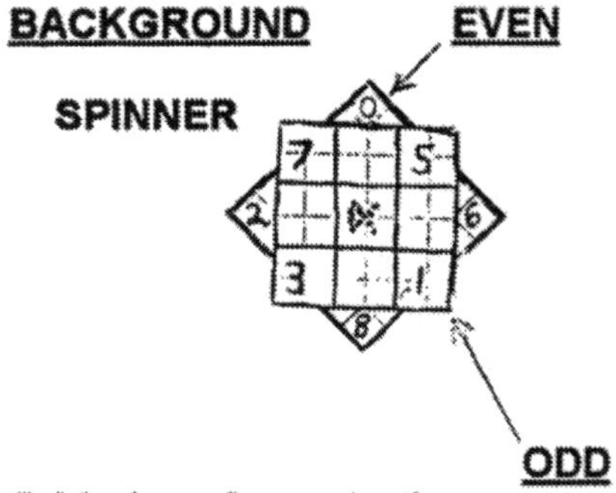

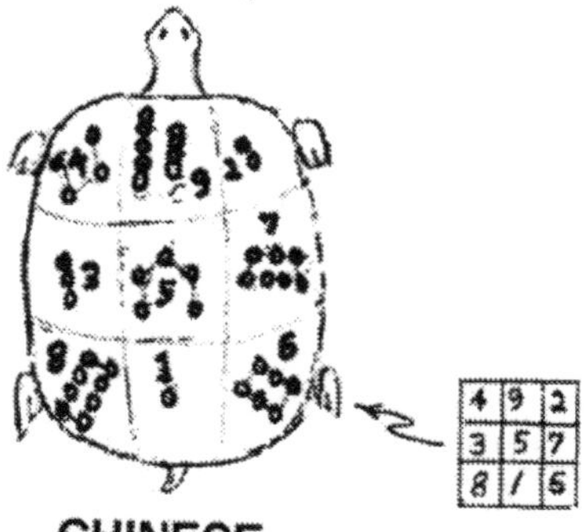

<u>BACKGROUND</u>

SPINNER

<u>EVEN</u>

<u>ODD</u>

CHINESE TURTLE

Odd plane forms to the even Hilbert space.

Compare the segments taken from the all *odd* isotropically *96* arranged momentum chart that is ejected from the pion scheme illustrated on page 21. Notice that the Hilbert orthogonal space background maintains the symmetrical field alignment during the "weak interaction" of the decay process.

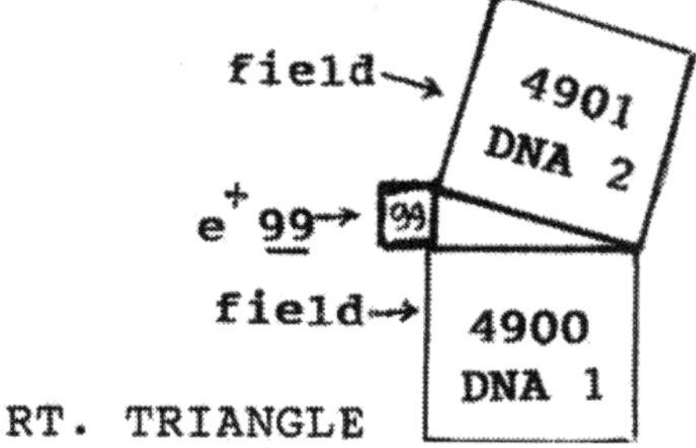

This chart, representing the mathematical and geometrical scheme of the electric helic field of a DNA molecule is *indisputable* evidence of the super positioning that occurs in the formation of DNA.

It stems from the efforts of the secret ancient Pythagorean Society in Greece. They realized that the *99 4900 4901* right triangle played a major part in some way in cosmic structure. Some considered it would reveal divine secrets. Pythagorean triples become increasingly more difficult to find, as the square of the number increases. The DNA fielding system is displayed on a large 4 ft. X 6 ft. chart. Send your request for additional data and/or cost of a CAP Chart or DNA chart to:

PENNICK SCIENTIFIC ENTERPRISES, P.O. Box 382, CARNEGIE, PA 15106-0382, or *pennscience@hotmail.com.* http://www.pennickscientificenterprises.com/about.html

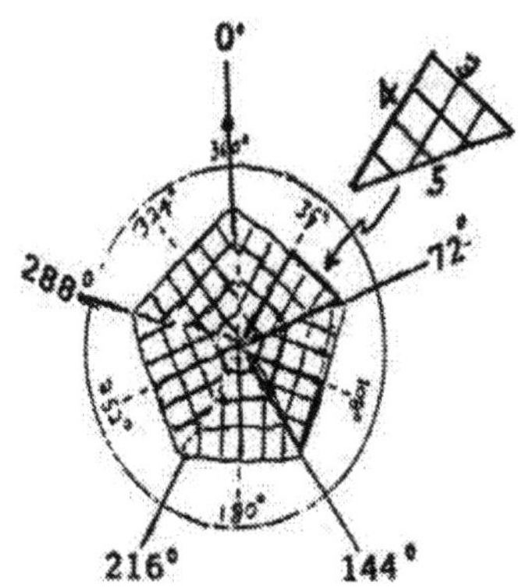

Doubting Thomas doubts that Hilbert orthogonal (all even) "averaged out" space is the stilled background in which odd numbered amounts unite to form the zero base of energy. So, let Thomas try to make a odd-even chart (as shown above) by INHERENT DNAputting the even plane upon the SPIDER MATH odd plane, instead of the way it 10 x $\sqrt{2}$ is structured above.

CONCERNING ODD AND EVEN AMOUNTS:

•Euclid stated that, "any three
The Pythagorean 3-4-5points in space can be
right triangle as $\sqrt{2}$, iscircumscribed into a flat (2D)
prevalent in thecircle.
mathematical function of•Minkowski stated that, "any
the low life spider web-4 points is space form an
spun constructionevent. (Minkowski-Einstein
4D theory)

CONCERNING CHARACTERISTICS ACQUIRED FROM PATTERNS THAT CROSS BETWEEN THE AMINO ACID MOLECULES IN DNA

Lower forms of life derive their characteristics from the patterns created by the 20/7 tightening of natures endeavor to square the circle. This endeavor is mathematically shown as displayed on the decimal expansion if Pi. Lower forms of life acquire their characteristics that stem from a minimum amount of decimal places of pi. Whereas, the intellectual characteristics acquired by humans are derived from 60 to 70 decimal places of pi (THI SIS THE *ONLY* PLACE where a complete set of ten digits, 0-1-2-3-4-5-6-7-8-9, can be found in the vast decimal expansion differentiation math scheme. In order to play with a full deck, it is necessary to acquire all of the 0 through 9 digits. 60 to 70 decimal places of pi is where humankind derive their *full-deck* intelligence. Well, maybe some of us play with a full deck? More about this transfer of decimal expanse of pi *across* DNA molecule is explained in the text book, "*The Price Principles of the Heisenberg Zone*".

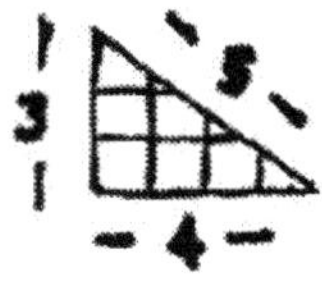

Before revealing the super positioning aspects of a Pythagoras right triangle scheme, it's necessary to add the 4-dimensional *mass* zero-point system to the 3X3 energy zero-point system, in order to complete Einstein's

Mass-Energy Equivalence postulate. The energy segment of his postulate is on page 75, as the zero-point *3X3* chart. The mass part of the mass-energy partnership is housed in a *4X4* chart that has 16 positions that compose the starting zero-point place of mass forms. The *3X3* (energy chart) along with the *4X4* mass chart, presents two of the three squares designated on the Pythagorean right triangle. The hypotenuse (side of right angled triangle opposite the 90 degree right angle), houses a 5X5 chart that is the zero-point the muon particle (later).

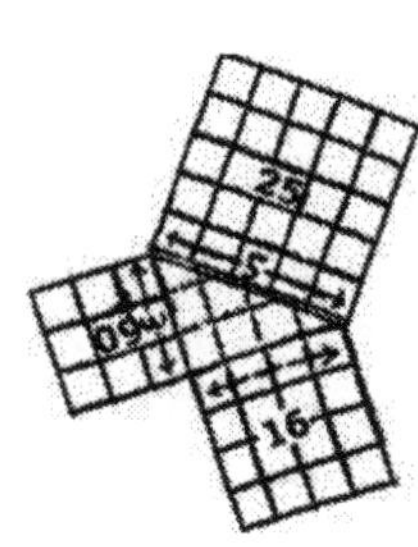

the 3-4-5 (squared) triangled shown at left, is Pythagoras Scheme. The 3-4-5 illustrated to the right, is the scheme. Einstein tried to conjure up a unified field theory which would super-position fields in order to account for the size of particles. The Phythagoras Scheme requires two different fields. My scheme is united, if it exists in a single field.

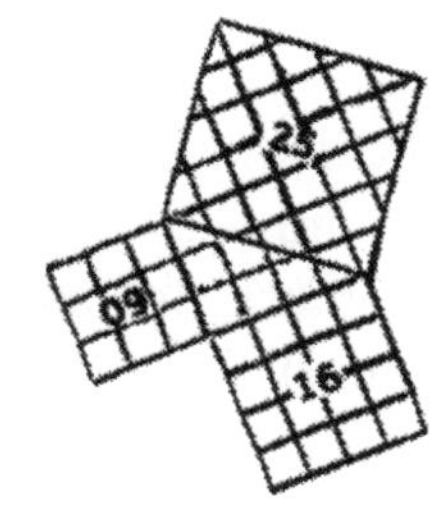

The data herein presented, is a brief summary of the relationship of energy-mass charting. A full explanation is related in the book called, "*The Price Principles of the Heisenberg Zone*". This scheme solves the Heisenberg $\Delta X \cong \Delta Y$ problem. It focuses it to $\Delta X \cong \Delta Y$.

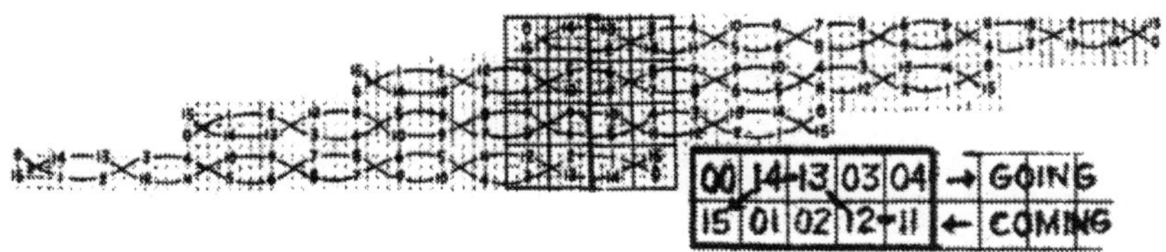

COMING-while-GOING RECIPROCATING MASS-ENERGY RELATIONSHIP

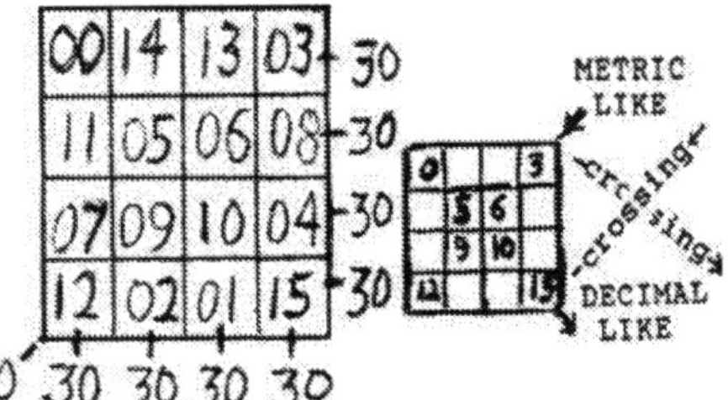

This *4X4* 16 potisional chart is the zero-point system from which subatomic particles are super-positioned from. This phenomena will be explained later.

Read more about the super positioning of energy via of the zero-point 3X3 chart on page 68, and also briefly study 4X4 charts that are displayed near the 3X3 chart in Sec 3 1 displayed on page 21.

The superpositioning of a 3-4-5 right triangle into the 9-16-25 (Pythagoras scheme) system happens at the onstart of combined energy-mass forms in the cosmos at zero-space. The 4X4 (16 positional) formed off of the base of the Pythagoras scheme is the system that is the zero-point of mass forms as protons and electrons. Euler, (or was it Euclid) stated that any 3 points, in space, can be circumscribed into a circle. Someone, (maybe me?) said 4 points, in space, can determine a square. Therefore space itself endeavors to "straighten" the curve of a circle (curvefitting). The square form is akin to Hilbert Orthogonal (all evened) space. 3 is odd, from which a circle can be derived --- 4 is even, from which a square can be structured. The following diagram is a graphed chart of the zero-point of mass. Bear with me, there is more to come.

A few schemes to familiarize the reader about 4-D zero-point that superpositions into mass forms are as:

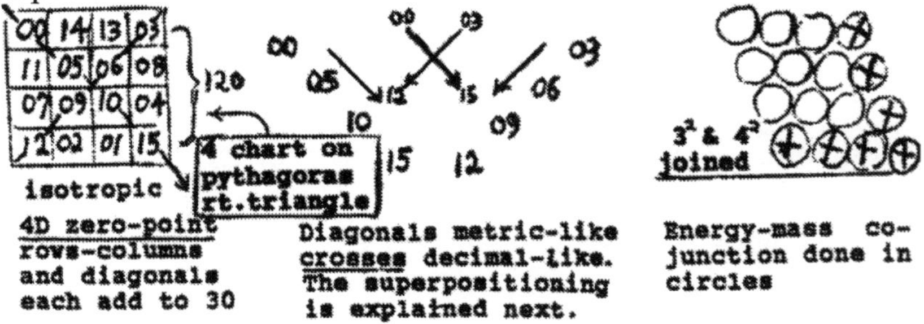

4D zero-point rows-columns and diagonals each add to 30

Diagonals metric-like crosses decimal-like. The superpositioning is explained next.

Energy-mass co-junction done in circles

FII – THE REASON THAT A 4X4 CHART IS THE ZERO-POINT OF MASS

The stable proton and its bonded co-partner (electron) together form the large field of current that houses a hydrogen atom. Particle accelerators produce particles named pi-mesons (pion). Pions are similar to protons except that pions are smaller and are derived from the result of high energy proton collisions. Pions undergo an exchange from an 8X8 sized field having 64 positions. 15 positions are ejected from 64 which allows 49 positions to remain. This pion scheme is displayed later in this The 15 positions make up the momentum that is able to exist in the 4X4 zero-point chart, from which all mass is super-positioned from. A few of the mathematical proofs that concern the 7+8 making up the the 15 (total momenta) of the 4X4 zero-point of mass that can super-position into larger forms of mass. Octonion arithmetic is included with the following diagrams that reveal the multiplication data that is involved in natures super-positioning schemes. See Bulletin of the American Mathematical Society, vol. 39 pg. 145.

see pion decay mode on page 21.

THE REASON WHY A 3X3 CHART DEPICTS THE ZERO-POINT OF ENERGY

Nature has a method that arranges the 3X3 isotropic chart so that it overcomes the Heisenberg Uncertainty Principle that does not permit the focusing of momentum into positions. The Maximum momentum that is in the 3X3 chart is limited to 8. The zero-point (initial system is space) of energy holds the 3X3 system to 8, which in algebratic form does not allow a procedure past 1,2,4,8. (The next number would be 16, which does not fit into a 3X3 charted scheme) 1,2,4,8 – that's it.

Follow down these diagrams in order to visualize the onstart of natures "initial" space transformation arranging schemes.

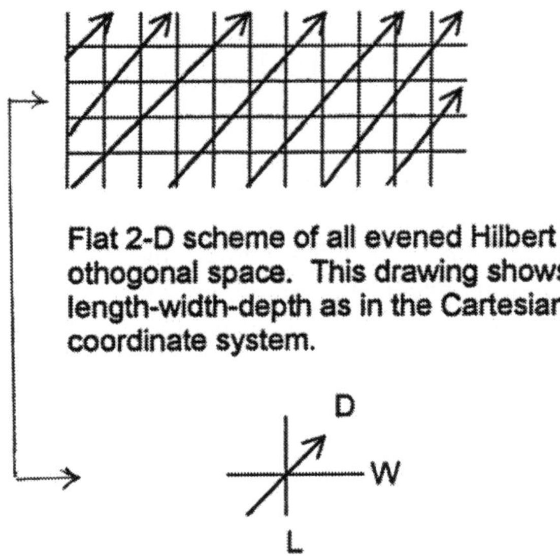

Flat 2-D scheme of all evened Hilbert othogonal space. This drawing shows length-width-depth as in the Cartesian coordinate system.

Albert Einstein used Hilbert space in his works as a background that mass fits into.

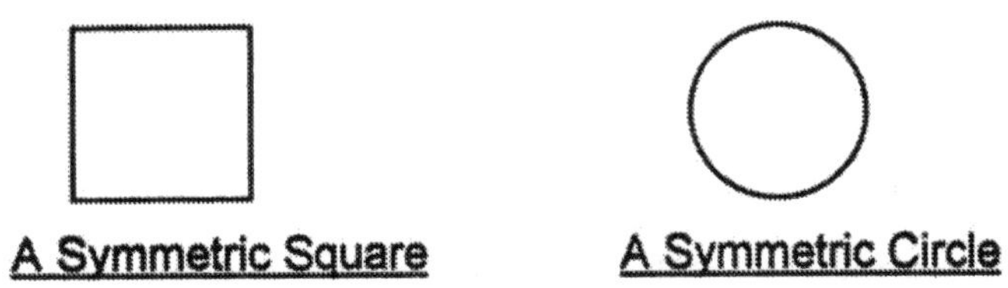

Both of these flat 2-D geometric forms are able to be "situated" into Hilbert orthogonal 3-D, space in a manner that contacts *depth*, as well as *W* & *L*. Both forms are closed systems. Euler or Uclid stated that *any* three points, in space, can be circumscribed into a circle. Minkowski – any 4 pts. in space constitutes an event (Einstein 4D space). In order for momentum to exist for a while, within and closed form system, the momentum must be arranged isotropically.

The arrangement of momentum that is in a system is isotropic if the system exhibits the *same* total amount when observed from any direction. The following diagrams illustrate the manner that nature employs in the very onstart of the formation of mass structures into space.

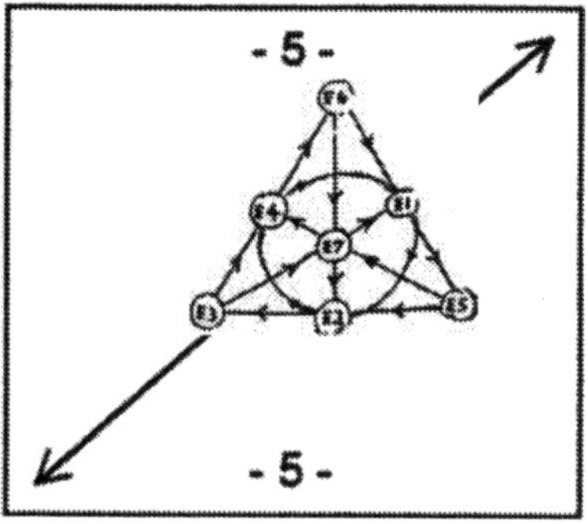

The ratio of the perimeter of a square compared to its diagonal, taken to the nearest fraction is 20 divided by 7. 20/7 = 2+6/7. Twenty divided by seven, when not taken to its nearest fraction is 20/7 =2.857142857142857 ---repeating to infinity. The square sized diagram shows the perimeter as having its 4 sides (of 5) which makes the perimeter 20. The diagonal is 7.

OCTONION MULTIPLIER

This system multiplies 8 dimensional numbers. Its isotropically arranged body multiplies, when following the arrowed lines (see octonion arithmetic). American Mathematical Society volume 39, page 145 (2002) by JOHN BAEZ.

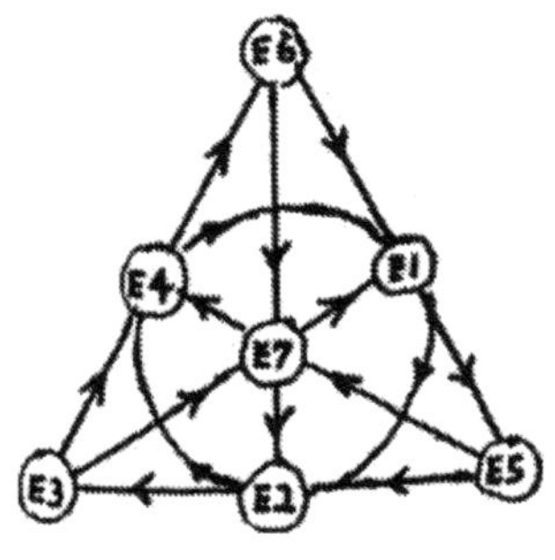

This triangular system plays a major part in the fundamental onstart that multiplies energy into

mass forms that superposition into higher forms. It is a form of curve-fitting that occurs at the weesmall area of zero-space. A release of force from the unbinding of the cosmos 20/7 gravity scheme. It assist in the expansion of zero-point energy and mass systems that reform to larger forms. see section II *plate* B

2 7/7 = 21 [7X2=14+7=21] 3 0/7=21 [7X3=21].
20/7=2.857142857142857142 --- infinite repeater →
22/7=3.142857142857142857 --- infinite repeater →
5.999999999999999999 ----------------------→

TRUE PI (not taken to the nearest fraction) is pi=3.1415926535 ---- continues onto infinity ---.

The 20/7 and 22/7 combined together add up to be 999999 →, which constitutes a full house of digits *0 through 9.* 9 is the maximum amount in the decimal 10 system. 0123456789 is *10* positions.

The radio of the circumference of a circle compared to its diagonal is 22/7, taken to its nearest fraction.

22/7=3.142857142857142 →
infinite repeater ---
pi= 3.1415926535 – infinite -----------→
decimal expanse –

The mathematical scheme of pi illustrates natures endeavor to square the circle. To square the circle means *making a square* that contains the same area *as in a circle*. As the decimal expansion of pi unravels, the difference becomes smaller, but never reaches a place where it squares a circle. It is this pi expanse system that is drawn across the amino acids in DNA molecules that understructures the characteristics of all life forms that exist over the ages. Animal and plant life-dinousaurs-ants-fish-birds-man amoebas-reptiles-hair color race-everything!

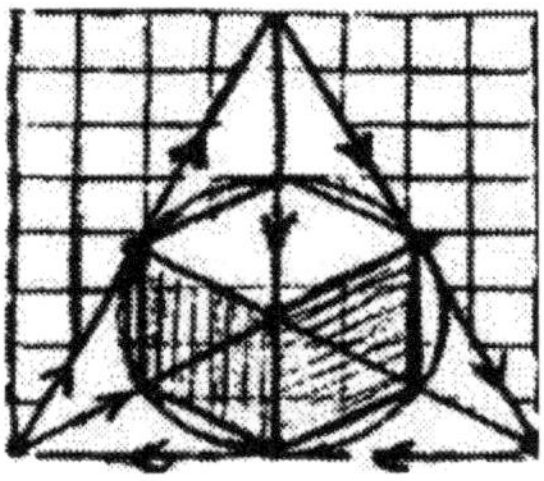

The space transformation
endeavor that squares the
circle for a brief time. It's
named FANO PLANE →Note: 3 pyramids are surrounding the Fano plane.
-- opposite spins –

Activated into
4-Dimensions

↓

↗

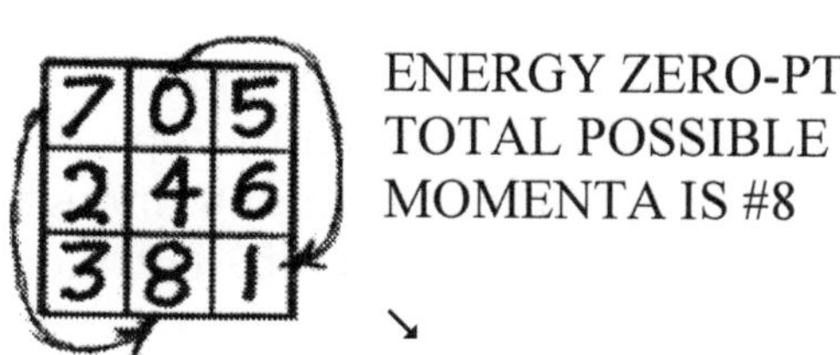

The front-runner that stipulates the formation of *all* mass forms.

YIN-YANG
Motion/stillness equalizer. This scheme attempts circle squaring. Like uncertainty. An ancient Chinese mathematician fine tuned the 22/7 pi, to be pi = 355/113.

The octonion multiplier scheme represents the initial manner of natures space transformation. It is the *forerunner* of the super positioning schemes that form all mass. It manipulates as the zero-point system of positions that make energy-mass jointure possible. It holds the circlesquaring endeavor into focus within its 8 multiplier system, balancing momentum and position for a brief while (21/7 = 3). Super mathematicians try to deal with the octonion multipliers symmetrical form. It limits the increase of 1,2,4,8 as the only *real* numbers. The next number after 8 would be 16, but 16 is not accepted by a "special algebra" as being the next term. This fact is why the octonion multiplier scheme is the ultimate onstarter base from which *all* mass is structured [entire solar system]. The atoms, listed along the diagonal of the CAP chart from hydrogen up to neon, all have their positive configuration *return* "through" the 10X10 field of the hydrogen atom. Oxygen having eight electrons, returns through #8, up the diagonal. Carbon, having six electrons returns through #6 etc., etc.

The atomic particle names a pi-meson (pion) exchanges the strong repelling forces that occur between protons housed together in the same atom. This exchange action is shown later in t his article. A pion is housed in an isotropic system having 8 rows of 8 positions (8X8 chart). It is a product of the octonion multiplier superpositioning action that sheds one row and one column, becoming an isotropically arranged 7X7 charted system containing 49 positions from 0 through 48. The 15 shedded positions (8&7) happens because the octonion multiplier stablizes its superpositioning in the manner shown above in the octonion multiplier illustration. The automuc particle names a proton, is superpositioned on up to a, isotropically arranged, 50X50 positional system, that continuously ejects (sheds) 99 positions (an electron), and becomes housed in a 49X49 isotropically arranged system. This phenomena is geometrically displayed next. Then after the proton display is the mathematical proof that describes why the electron is ejected in 99 positions as the momentum total housed in a 10X10 system that contains 100 positions, 0 through 99. [See electron locus displayed on page 9].

An electron is unique in its 99 positional scheme due to the difference that is involved in cosmic mathematical alignment of the squares of numbers as [See the square scale on the following page]. Later in this article a display of the 49X49 (2401 positions) is shown, done in circles, as being the scheme of a proton. Free protons are rare! Protons immediately connect with an electron as its co-partner. This connection is accomplished by a 50X50 (2500 positions) isotropically arranged system. The 50^2 system cannot exist in the same field as does the stable 49^2 system because a 50^2 has too many positions to fit into the same field as a 49^2 proton. The reason it can't fit is due to the fact that the initial position is zero. 0 adds another position. EXAMPLE; 0 plus 99 = 100 positions. 100 plus 0 = 101 positions. 50^2 is past ½ way in the field area.

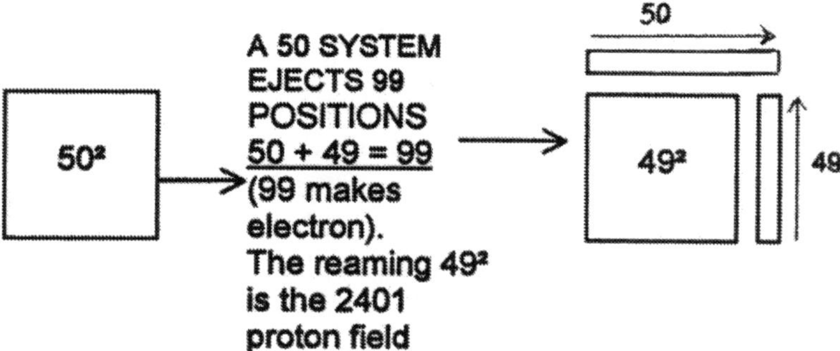

The mathematical scale that proves the stable 99 system is depicted as:

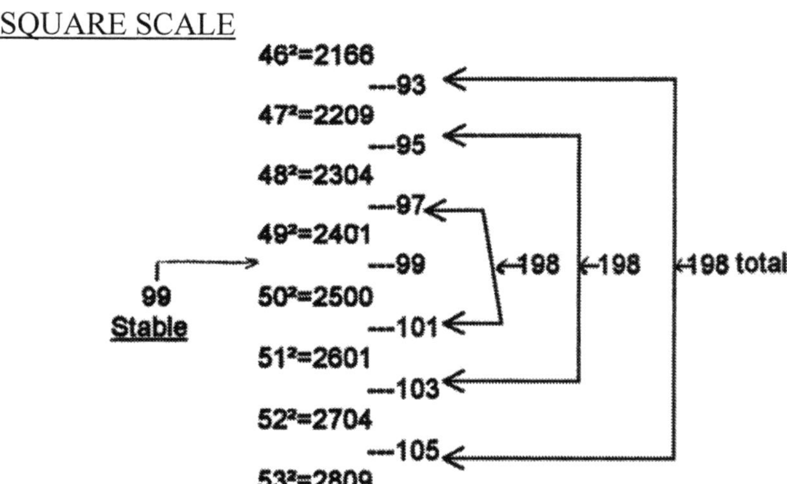

above *square scale* is a small segment of a larger scale. It is the part that shows the stable 99 and the proton field action. 99 is the difference between 46^2 and 47^2. 95 is the difference of 47^2 and 48^2 etc. etc.

Energy induced into the momenta of isotropically arranged systems causes the superpositioning that makes up larger isotropic systems. The maximum reach of superpositioning engulfs the solar system. The CAP chart displays ten atoms hydrogen to neon. Each position up the diagonal represents a superposition. The field of the CAP chart itself, when multiplied by 99, becomes the starter system of the double stranded helix DNA molecule. The DNA chart is huge. It is supplied with the book called "*The Price Principles of the Heisenberg Zone*".

Mapping the complete human genome was a major achievement, but, it is like having a road map without the towns and cities displayed on it. The DNA chart that is supplied with the book "*The Price Principles of the Heisenberg Zone*", portrays the underlying mathematical and geometrical scheme that outlines the current flowing between the helic strands that cross the field that houses the atoms that make up the amino acid sets. The chart replaces the currently used ball and stick (tinkertoy like scheme), that depict atoms and electrical bonding current with the structural details of the atoms nucleus and the momentum/position changing aspects of the electron bonding happening between the atoms in the amino groups. It details the superpositioining stages that determine the formation of characteristics --- species-race-eye color-size etc., etc., etc. Its like adding the cities and road signs on a near blank map.

Douglas M. Price

GLOSSARY

Angular momentum
Momenta that curves from a straight locus. Rotates etc.

Apprehensive space
Vacant space is apprehensive to all probes. It is the foremost and most accessible and approachable phenomena.

Binding force
A tightening of momenta toward the center of a system. An increasingly stronger wrapping force. The decrease in size that happens in the decimal expansion field of pi which is bound in protons in atoms. (see strong force)

Circle squaring
The endeavor that the cosmos undergoes in its attempt to straighten out curved form. The differentiation of the decimal expansion of pi.

Decay mode
The manner that sub-atomic particles release momenta and form into smaller particles of mass.

Electron (bound)
Resides outside of the positive field area of a proton. It is attached to the proton by positive+ and negative- connected current that exist between a proton & electron.

Force (centrifugal)
A curve-fitting 20/7 & 22/7 force that causes objects to eject outward in a straight line from a circular swinging motion. It is a form of circle-squaring action.

Free protons
Free protons are *rare*. They immediately connect up to an electron. This is accomplished by the co-juncture of negative- and positive+

current that is established between the positive field+ of the proton and the negataive field- of the electron.

Hilbert space

Hilbert orthogonal space implies that free space is averaged out—into an all evened form. (no odd).

Intrinsic

The meaning of intrinsic that relates to sub-atomic data is inward like incoming current. Extrinsic would refer to outgoing current. (coming---going)

Isotropic

A system is considered to be in isotropical balance if its momenta, that is occurring throughout the positions of the system, add up to the same amount relative to an observer from all surrounding directions. (a symmetry thing).

Neutrino

A basic minute, isotropically arranged system that that consist of all even or all odd amounts of momentum. When a system that consists of all even (or all odd) momentea, it exists only as a *potential* system in stilled space. In order for a system to aquire a "presence" in zero-space, even requires odd to activate the evened Hilbert space into action. In Hilbert space, math sets up as 2+2=4 0+4=4 and is undetectable as such. but when 3+1=4 happens, the system is able to have an existence as a minute wave like 0 2 4 1 3 this set occupies 5 positions set in two planes.

Neutron

As isotropically fielded particle. It is thee system that neutralizes the 137 strong nuclear force that repels between the positive+ charged protons that are together in the nucleus system of an atom. The repelling strong force is exchanged via .a pi-meson particle that transfers the force, like a go-between that transfers opposite currents.

Octonion multiplier

An octonion multiplier is the basic system that exchanges momentum & position making it possible to "SUPER-POSITION" mass forms into large mass forms. (super positioning).

Particle
Refers to a particle or sub-atomic particle of mass.

Position
A given area in Hilbert space where momenta is housed.

Positron
An atomic particle that is equal and opposite of its anti-particle which is an electron Positrons carry a positive+ charge, whereas electrons carry a negative- charge.

POTENTIAL ENERGY
Energy that is already set in a "potential" field that consists of all evened amounts of momentum. (Hilbert space still field. When the stilled field is activated by motion even and odd join to motivate action, work etc.

Potential field
Consists of all evened(averaged out) Hilbert orthogonal space. It is the area in space position that becomes activated by energy that can house mass etc.

Proton
A particle of mass that resides in the nucleus (core) of an atom. Free protons are rare because they have current ties to their co-partner electrons.

Presence
A bit of time lapse caused by a *coming* (from future) type of field that cojunctions with a *going* type of field , which creates a presence that consists of a coming/while/going system. The proton, and its counterpart electron, maintain a stable presence due to the down quark field coming from the future, meshing with up going quark field.which results in the formation of the stable presence of a

proton. The electron is housed in a stable coming/while going jointure of opposite fields that stabilize a presence of time. lasts).

Planck energy unit
A fundamental, isotropically arranged, system that activates action into a system. It is a symmetric tensor of the ninth rank, having eighty one vector-scaler (even-odd) components in its 9X9 system of 81 positions.

Quarkery
The initial field system that combines with another quark system to form a particle of mass. Up quarks—down quarks strange quarks etc. An electron is structured with a 33 positional quark & a 66 positional quark in its 99 system.

Spin
The coming/while/going scheme of the locus of momentum in a particle adds a twist that exerts opposite force issued from the center of the particle.

Stable particle
A system that contains isotropically arranged momentum. The only two systems that maintain there stability are electrons and protons. A neutron particle has a ½ life lasting about 12minutes.

Symmetry
A system having balance, equalized thru out. A system that has isotropically arranged momentum in its positional set. Equally balanced from all directions, relative to observer

Space transformation
The exchange of the momentum from one area of space into another. Einstein—X + Y + Z = X' + Y' + Z' .

Weak force
The small force that occurs during the coming apart decay mode of sub-atomic particles. Weak interaction process.

Zero-point

The initially wee-small system that a quantum particle of mass is super-positioned from to expand into its larger form. At Princeton University (1941) Einstein-Bergman and Bargmann worked on a theory that electrons being derived as point-particles in space.

Zero-space

Like the *non-existent* area that lies beyond the reach of starlight from the Universe and anywhere else that makes up an area of un-occupied space. Zero space.

Zero-zone

Refers to the initial area where the onstart of mass forms from. The realm of the Heisenberg Zone is small like the size of electrons (16-^{m}). Where momentum and position can't be focused without being housed in a "coming/while/going" isotropically arranged system.

About the Author

A general contractor in all fields of construction.

He became interested in atomic research shortly after world war two. His high school math teacher worked at Oak Ridge on the Manhatten Project during her summer vacations. When she presented the numberline (-4 –3 –2 –1 0 +1 +2 +3 +4) as a mathematical axiom, Douglas disagreed. He knew that the numberline worked out when counting bricks etc., but when something forms from zero space, the numberline method is non-functionable. This began his endeavor to unravel the unknown aspects concerning to structure of sub-atomic mass. The wee-small systems that come together changing momentum and positions that form mass in zero-space.

During a lifetime of research, he accomplished his endeavor to chart the momentum changing position that occurs in sub-atomic particles. He did not quit until he completed Albert Einsteins dream of a grand unified field theory and charting the inner-workings of DNA molecules. At the present time, he is doing even more important research (revealed in this book).

www.ingramcontent.com/pod-product-compliance
Ingram Content Group UK Ltd.
Pitfield, Milton Keynes, MK11 3LW, UK
UKHW041938190726
13854UKWH00004B/1651

9 781410 749086